THE SPIRIT OF THOREAU

Elevating Ourselves

The Spirit of Thoreau

SPONSORED BY THE THOREAU SOCIETY

Wesley T. Mott, Series Editor

✦

Uncommon Learning:
Thoreau on Education

Material Faith:
Thoreau on Science

Elevating Ourselves:
Thoreau on Mountains

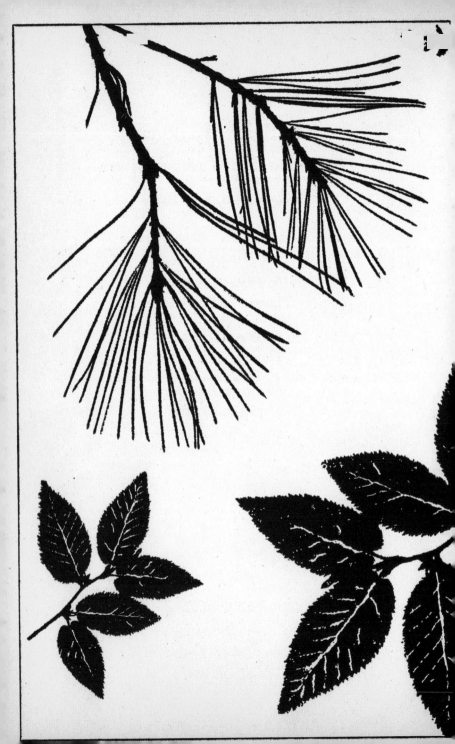

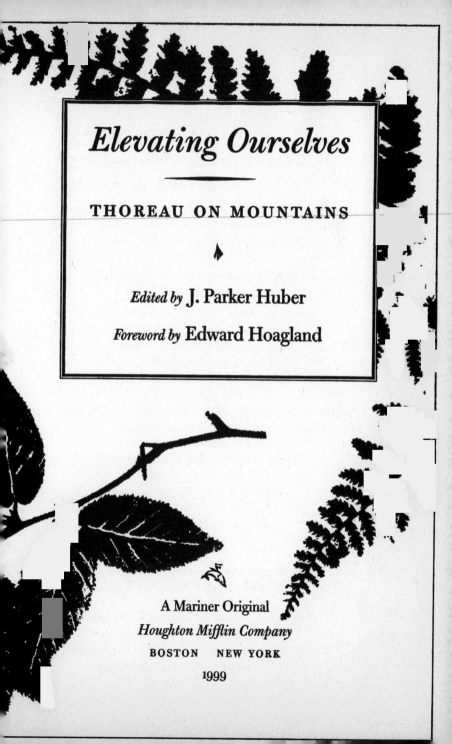

Elevating Ourselves

THOREAU ON MOUNTAINS

Edited by J. Parker Huber

Foreword by Edward Hoagland

A Mariner Original
Houghton Mifflin Company
BOSTON   NEW YORK
1999

For information about permission to reproduce selections
from this book, write to Permissions, Houghton Mifflin Company,
215 Park Avenue South, New York, New York 10003.

Library of Congress Cataloging-in-Publication Data

Thoreau, Henry David, 1817–1862.
Elevating ourselves : Thoreau on mountains / edited by
J. Parker Huber ; foreword by Edward Hoagland.
    p.   cm. — (The spirit of Thoreau)
"A Mariner original."
Includes bibliographical references (p.).
ISBN 0-395-94799-5
1. Thoreau, Henry David, 1817–1862 Quotations.   2. Mountains
Quotations, maxims, etc.   I. Huber, J. Parker   II. Title.
III. Series: Thoreau,
Henry David, 1817–1862. Spirit of Thoreau.
PS3042.H74   1999
818'.309—dc21   99-15172   CIP

Book design by Anne Chalmers
Type: Bulmer (Monotype)

Printed in the United States of America
QBP 10 9 8 7 6 5 4 3 2 1

Selections from volumes 1–5 of Thoreau's *Journal*
are from *The Writings of Henry David Thoreau*, Witherell, E., ed.
Copyright © 1972 by Princeton University Press. Reprinted
by permission of Princeton University Press.

On the tops of mountains, as everywhere
to hopeful souls, it is always morning.
— The first version of *Walden*

FOR CINDY PAULEY
... wherever she came it was spring.
— *Walden*

FOR ALBERT BUSSEWITZ,
WALTER HARDING,
EDWIN AND NELLIE TEALE
So our human life but dies down to its root,
and still puts forth its green blade to eternity.
— *Walden*

FOR ANN HAYMOND ZWINGER
Humility like darkness reveals
the heavenly lights.
— *Walden*

Contents

———

Foreword

SEPTEMBER 7, 1852, "was a colossal day for Thoreau," says J. Parker Huber, the editor of this invigorating compilation, who retraced Thoreau's seventeen-mile walk from Peterborough, New Hampshire, up Mount Monadnock (he had botanized half a dozen different kinds of berries just on the summit), and then down the other side to the village of Troy, where a railroad station afforded Thoreau a quick return to Concord by late afternoon, *with the plants of the mt fresh in my hat.* On jaunts like this, Thoreau was not the hectoring social critic some readers still object to in the early sections of *Walden.* He was on a lark, sauntering in the mountains and "elevating" himself, replenishing his "artesian" wellsprings, and jotting in his journal trout-trembly observations not especially meant for publication. His mind is lighter and not in the evangelical mode — not correcting a wage-slaving, slave-holding nation that was obsessed with its conquests in the Mexican War and the Gold Rush in California. Even looking down from Monadnock, or from Mount Lafayette, way north in the White Mountains, upon haphazardly stripped forests foretelling overdevelopment, he calls the land "leopard-spotted," a romantic metaphor for clear-cuts. And he is usually with a friend —Harrison Blake or Ellery Channing, Edward Hoar or George Thatcher—not at all the grumpy solitary of legend.

Thoreau had many facets. He was perhaps our best essayist and finest literary stylist, also a first-rate field biologist, and an inspiration to Gandhi and other avatars of later civil disobedience. But what he mainly sought to do in his short life—as exemplified by his masterpiece, *Walden*—was show us how to live. In other words, he didn't aspire to be a one-of-a-kind, heroic type, a proto-conqueror of Everests, or a Charles Lindbergh sort, winging toward fame in the heights of the sky, nor even a figure resembling his posthumous admirer John Muir, who discovered glacier fields far off and alone in Alaska, after exploring the High Sierras. Instead, Thoreau wished to demonstrate how practically anybody, even nowadays, may still choose to live independently amid the back lots of a country town—which is a philosophical, not a physical, challenge.

The twenty ascents Mr. Huber has distilled from Thoreau's correspondence, his diary entries, and magazine and book fragments are trips such as we all might make. Climbing Wachusett, near Worcester, Massachusetts, Thoreau reached an altitude of only 2006 feet, and four of his other mountains (Kineo, on Moosehead Lake in Maine, plus Wantastiquet, Uncanoonuc, and Fall, in New Hampshire) are under two thousand feet. That is really the point, as you read Thoreau. He went nowhere that we can't follow. In the Berkshires, the Hoosacs, the Catskills, as well as the stiffer, celebrated pitches of

Mounts Washington and Katahdin, Mr. Huber repeats all of his walks—sometimes coming out of the woods "no less elated than Balboa at the Pacific," he says—while painstakingly retrieving the truth about disputed routes and campsites, and recapitulating what has happened to these numerous places in the years since. Several of the bogs of Monadnock, for instance, are "still living," as he felicitously puts it.

Elation is the name of the game here. Thoreau is not trying to alter the onrush of Capitalism. Waking one night on Kineo, he delights in discovering a phosphorescent log that "made a believer of me more than before. I believed that the woods were not tenantless, but choke-full of honest spirits as good as myself any day,—not an empty chamber, in which chemistry was left to work alone, but an inhabited house,—and for a few moments I enjoyed fellowship with them." Yet Monadnock, 3165 feet high, in southern New Hampshire, on visits during 1844, 1852, 1858, and 1860, may have provided Henry with his happiest intimacy with any mountaintop. The intricacy of its rocks, brush, and berries—geology right next to his face—and clouds that coalesced as scuds or parasols, then gradually dissipated overhead, or over Saddleback (now Greylock) Mountain, sixty miles away, where Massachusetts's Berkshires meet the Green Mountains of Vermont, pleased his fancy and stretched his powers of speculation.

On Lafayette, in the dwarf fir thickets near Eagle Lake, deep reddish-purple twinflowers were "altogether the fairest mountain flowers I saw." Of Mount Washington, which he had first ascended with his beloved brother, John, in 1839, he surely spoke for all the heights: the top "should not be private property; it should be left unappropriated for modesty and reverence's sake, or if only to suggest that earth has higher uses than we put her to." The two of them had walked eighty-two miles in five days, Mr. Huber estimates. Nineteen years afterward, Henry returned for a longer tour with Edward Hoar—though rather careless, as ever, with fire. Despite having accidentally burned three hundred acres of Concord's forest in 1844, during this one year, 1858, he was twice in on close calls with a wildfire: above Tuckerman's Ravine and on Monadnock.

You will notice, however, that there are no *feats* in Thoreau. John Muir was ten times the hiker in the treks he completed and the peaks (though gently) he scaled. Yet that's why we need them both, transcendent on different coasts and devout in quite different realms. Thoreau was not a first-ascent man, with childhood demons like Muir's to expunge, nor an archetype conservationist with that underdog drive that finally exalted Muir in middle age to lasting achievements, both political and literary, that were uniquely his own. But if Thoreau didn't save the Yosemite Valley and found the Sierra Club, his corrective impulses ramified wider. And at a point when so many people now feel hardly more of a link to outdoor nature than as if it were sort of a dewclaw, Thoreau can remain an exemplar for ordinary folk who want to climb modest eminences near their homes —like his Pack mountains in southern New Hampshire, and North and South mountains, near the Hudson River in the Catskills. We can all be with him on a Sunday walk or an overnighter as long as we recognize, too, there in our tent, that Thoreau was attempting to recalibrate civilization, as well as simply sleep out. Nature was to stand in a central position, not as a vestige, or pet, or vehicle for our narcissism, such as it is when we scramble uphill mainly for an aerobic exercise to condition the body, like a stint of sidewalk jogging or pedaling gymnasium machinery.

Up in the sopping fog, or camped in a col halfway down again after climbing the rise, Thoreau was exuberant in his humility — not demeaning the mountain as so many outdoorsmen do in their mind's eye by considering it an obstacle to be inventoried and topped ("done that!" — which Muir, incidentally, despite his tireless athleticism, never did). And when the rainy fog cleared, Thoreau loved to look up just as much as down. Upward and out. Elevation was a blessing, not a conquest. "Only that day dawns to which we are awake," he had written, in closing *Walden*. And rather like reconstituting a withered limb, we at this new millennium must learn to admire each mountain for its complex spirit and self, not the illusion that the world lies at our feet.

EDWARD HOAGLAND

Elevating Ourselves

Introduction

We yearn to see the Mts *daily*

THOREAU went awandering for the sake of his soul: "I cannot preserve my health and spirits, unless I spend four hours a day at least,—and it is commonly more than that,—sauntering through the woods and over the hills and fields, absolutely free from all worldly engagements."

This was his daily ritual, his perennial practice. He often walked to higher ground, counting his blessings: "For hills—Nagog famous for huckleberries where I have seen hundreds of bushels at once—Nashoba—of Indian memory—from which you can see Uncanunuc Mt well—Strawberry hill—from which you glimpse Nagog Pond—Annursnuck—Ponkawtasset—Balls hill Fair Haven—Round—Goodman's—Willis's Nobscot ...by the Sudbury inn v poetry Turnpike hill—Lincoln Hill—Bare hill Mt Tabor, Pine hill, Prospect hill Nawshawtuct. Wind mill Hill. &c." His hilly communions are rendered fully in his *Journal*.

Thoreau climbed "to know what a world he inhabits." On 12 September 1851, he went to Flint's Pond "for the sake of the *Mt* view from the hill beyond looking over Concord....It is worth the while to see the *Mts* in the horizon once a day." He went to "take pleasure in beholding the form of a mountain in the horizon." In all seasons, he went for visual nourishment: "The hills seen from Fair Haven Pond make a wholly new landscape. Covered with snow & yellowish green or brown pines & shrub oaks they look higher & more massive. Their white mantle relates them to the clouds in the horizon & to the sky. Perchance what is light colored looks loftier than what is dark."

The way he looked at landscape was shaped in part by the English traveler William Gilpin, although he said that Gilpin "fails to show why roughness is essential to the picturesque, because he does not go beneath the surface." After following Gilpin through Wales, Cumberland, and Westmoreland, Thoreau was dismayed that he never reached a mountaintop.

Thoreau transformed the conventions of travel, making the search for a beautiful landscape a spiritual journey. "My profession is to be always on the alert to find God in nature," he said. He wished to be "elevated for an instant upon Pisgah," for then the world became "living & divine." "*Mts.* thus seen are worthy of worship." They "elevated and etherialized" him. This was the real purpose of transcendental travel, to quest for "your own higher latitudes."

From time to time I overlook the promised land....

Besides enjoying the view of distant mountains, Thoreau went to them. He began at age twenty-two, in late summer 1839, with Washington, and ended twenty-one years and twenty ascents later, in 1860, with Monadnock. Interestingly, these two mountains, and Wachusett, were the ones he returned to more than once.

Thoreau's Climbs

1. 10 September 1839 Washington, New Hampshire 6288'
2. 20 July 1842 Wachusett, Massachusetts 2006'
3. July 1844 Monadnock, New Hampshire 3165'
4. July 1844 Hoosac Range, Whitcomb Summit,
   Massachusetts 2173'
5. July 1844 Greylock, Massachusetts 3491'
6. July 1844 Catskills, New York 2200'
7. 7–8 September 1846 Katahdin, Maine 5267'
8. 5 September 1848 Uncanoonuc, New Hampshire 1329'
9. 6 September 1852 Temple Mountain, Whitcomb Peak,
   New Hampshire 1710'
10. 6 September 1852 Pack Monadnock, New
    Hampshire 2286'
11. 7 September 1852 Monadnock, New Hampshire 3165'
12. 19–20 October 1854 Wachusett, Massachusetts 2006'
13. 9 September 1856 Wantastiquet, New Hampshire 1351'
14. 10 September 1856 Fall, New Hampshire 1115'
15. 24 July 1857 Kineo, Maine 1806'
16. 2–4 June 1858 Monadnock, New Hampshire 3165'
17. 5 July 1858 Red Hill, New Hampshire 2029'
18. 7–8 July 1858 Washington, New Hampshire 6288'
19. 14–15 July 1858 Lafayette, New Hampshire 5260'
20. 4–9 August 1860 Monadnock, New Hampshire 3165'

His highest ground was Washington. Thoreau's elevations, taken from John Hayward's *New England Gazetteer* (1839), have changed owing to natural forces — the "work up, or work down" of the planet that Thoreau sensed on Katahdin—and owing to a new elevation reference system and more sophisticated technology; in 1997, for example, computer-satellite calculations augmented Vermont's Mount Mansfield by two feet, to 4395'. For my purposes, I have limited mountains to those of 1000' or more, though names are given for lower lands, such as Mount Ararat, a shape-shifting Cape Cod sand dune (100'?). Thoreau climbed Ararat twice, its ferocious winds reminding him of Mount Washington.

He made one ascent outside the United States. Mount Royal (763'), around which the city of Montreal grew, is one of several eminences between the Appalachians and the Laurentians called the Monteregian Hills. Thoreau's two-mile approach on 2 October 1850 — almost the 215th anniversary of Jacques Cartier's baptismal climb of 3 October 1635, of which Thoreau knew—was typical: "going across lots in spite of numerous signs threatening the severest penalties to trespassers.... From the mountain-top we [his friend the poet Ellery Channing accompanied him] got a view of the whole city; the flat, fertile, extensive island; the noble sea of the St. Lawrence swelling into lakes; the mountains about St. Hyacinth, and in Vermont and New York; and the mouth of the Ottawa in the west." This summit now stands in Mount Royal Park, designed by Frederick Law Olmsted. You can walk to the summit on the Olmsted Trail.

Though his Canadian walking tour included mountains, his focus was on water tumbling into the St. Lawrence River: the falls at Montmorency, St. Anne, La Puce, and Chaudière. On his three-mile walk up the east side of the St. Anne River to its falls, he climbed a steep hill and came close to Mont St. Anne (2625'), rising west of the falls. "I just remembered the wildness of St Anne's," he wrote the next summer, "that's the ultima Thule of wildness to me." It was also his farthest north.

Thoreau did not climb faraway mountains. The West went untouched. Not going below the Mason-Dixon Line, he missed seeing the Southern Appalachians, let alone walking through them, as John Muir did in 1867. Even within his ambit there were voids. He never explored the Green Mountains of Vermont, though his train crossed them coming and going to Canada in autumn 1850 and returning from Minnesota in the summer of 1861, and he saw them from Monadnock in southwestern New Hampshire. Vermont's pinnacle, Mansfield, climbed by Emerson in his sixty-fifth year with his younger daughter Ellen in the summer of 1868, and by John Muir in horse and carriage in an autumn rain and snowstorm in 1898, eluded him. In fact, he did not climb any Vermont mountains. When Emerson and other philosophers camped at Follansbee Pond in the Adirondacks of New York in August 1858, they hunted and fished instead of elevating themselves. Thoreau, who declined to join them, took in the Adirondacks from his train window when he approached Lake Champlain in 1850, and traveled north of them in July 1861. Had he been able to deliver his address on John Brown in North Elba on 4 July 1860, he would have been in their midst.

Of course, Thoreau was aware of many more mountains than he experienced. Reading, observing, and conversing with friends introduced him to others.

"The richest prospect in New England, and not improbably in the United States" was Timothy Dwight's assessment of what he beheld from Mount Holyoke in the autumn of 1810. Born in 1752 on the other side of the Connecticut River from Mount Holyoke, in Northampton, Massachusetts, Dwight became a Congregationalist minister and served as president of Yale for twenty-one years. He traveled extensively throughout the Northeast — about eighteen thousand miles, he estimated — recording his observations in letters. He even visited Concord, though not Walden. He was as keen as Thoreau on a good view, even celebrating that from the statehouse in Boston. His dying wish to have his history published resulted in *Travels in New England and New York,* which Thoreau read. Before mountain climbing became fashionable in America, Dwight went to the heights Thoreau would attain. In 1799 on horseback, he followed a cart road to the top of Saddle Mountain (Greylock) in northwestern Massachusetts, scaling a tree for the panorama. In 1813 he rode his horse up Red Mountain (now Red Hill), north of New Hampshire's Lake Winnipesaukee. In 1815 he went up what was then called the Catskills, now their eastern escarpment, partly on foot, partly on horseback. Others he witnessed: Lafayette, which he named Wentworth for a former New Hampshire governor; Washington, "the loftiest elevation in the United States"; and Wachusett, "a single eminence, of an obtuse, conical figure." Monadnock he saw only from afar. Of those in the central wilderness of Maine, then part of Massachusetts, he knew nothing.

Edward Hitchcock (1793–1864), professor of chemistry and natural history and president of Amherst College, agreed with Dwight that the Holyoke prospect was supreme. Thoreau read his *Report on the Geology, Mineralogy, Botany, and Zoology of Massachusetts* and may have seen the "View from Mount Holyoke" which Orra White Hitchcock drew to accompany her husband's book. Thoreau also read his *Final Report on the Geology of Massachusetts*, which included a panorama of the Holyoke Range. Mount Hitchcock of the Holyoke Range honors him, as well as the glacial lake that once extended northward from Glastonbury, Connecticut, and surrounded this mountain.

Emerson may have informed Thoreau about Mount Holyoke, too. In late August 1823, the twenty-year-old Emerson walked from Roxbury to Amherst, from where he went by chaise to Holyoke's summit. Unlike Emerson, who returned to Amherst several times to lecture, Thoreau never made it a destination. Emily Dickinson, who met Emerson in Amherst and had read *Walden*, loved this mountain range. Thoreau may also have seen or read about Thomas Cole's *View from Mount Holyoke, Northampton, Massachusetts, after a Thunderstorm*.

Surely he'd have heard of Margaret Fuller's adventures. On her 1844 autumn holiday with Caroline Sturgis in Fishkill Landing (now part of Beacon), New York, on the Hudson River, though still writing *Woman in the Nineteenth Century*, Fuller was free to ramble. "All the fine weather I have passed in the mountain passes, along the mountain brooks or the river," she wrote her brother, Richard Fuller, on 23 November. Curiously, these places are nameless in her letters, but a glance at the map reveals that just south of her lay the Hudson Highlands, with North and South Beacon mountains. She even took her visiting brother-in-law, Ellery Channing, into the mountains — in July he had been with Thoreau in the Berkshires and Catskills. Later, Thoreau must have read in her dispatch, on the front page of the 13 November 1846 *New York Daily Tribune,* of her ascent of Ben Lomond from the Rowardennan Inn on Loch Lomond, Scotland, with Marcus Spring but without guides, and of her being lost and spending the night with the mountain.

> *How cheering & glorious any landscape*
> *viewed from an eminence!*

Before you set out for these mountains, reflect upon Thoreau's advice to travel wisely — that is, pay full and kind attention to your means and purpose for movement.

For his 1839 White Mountain trip, Thoreau traveled by boat, stage, and on foot. In 1846 he used train, steamer, horse and buggy, batteau, and his legs to attain Katahdin. He approached Mount Kineo by canoe on Moosehead Lake in 1857. His 1858 carriage ride to Mount Washington caused him to chafe: "It is far more independent to travel on foot. You have to sacrifice so much to the horse. You cannot choose the most agreeable places in which to spend the noon, commanding the finest views, because commonly there is no water there, or you cannot get there with your horse."

So if you would travel as Thoreau did, consider foremost those places that can be earned by your own power. Revitalize the nineteenth-century walking tour. Try going by public transportation or combining train rides and walking and bicycling. My weekly bicycle (April to November) and train (other months) trips from Brattleboro to the Frost Library of Amherst College, where I read many books about Thoreau, have been a joy.

Embrace new ways of being with mountains. Make pilgrimages to them. Pray and meditate in these sanctuaries. Engage your heart with their elements. Walk around them. An ancient Hindu worship ritual, *pradak-shina,* was first performed in the United States by the poet Gary Snyder and others, who walked around Tamalpais, north of San Francisco, in 1965, and in New England by my walking around Monadnock in 1985. Circumambulating reveals much about inner and outer nature that the determinedly linear pursuit of a peak does not; it means creating your own way, which Thoreau encouraged—he enjoyed perambulating the bounds of his town—thereby exciting a spirit of inquiry.

Finally, relish staying at home, exploring your own terrain, taking a sabbatical from travel, leaving the wild in wilderness. Thoreau realized that he could be in touch with what was most meaningful anywhere. After many sojourns near and far, he held dearest one Concord shrine. "In all my rambles I have seen no landscape which can make me forget Fair Haven. I still sit on its Cliff in a new spring day & look over the awakening woods & the river & hear the new birds sing with the same delight as ever—it is as sweet a mystery to me as ever what this world is."

*The landscape lies far and fair within, and
the deepest thinker is the farthest traveled.*

Wachusett

19–22 JULY 1842

On 19 July at 4:45 A.M., Thoreau departed Concord on
foot with eighteen-year-old Richard Fuller, whom he was
preparing for Harvard, for "the blue wall which bounded
the western horizon," 29 miles due west. From his home
in Groton, at 108 Pleasant Street, Fuller could see Wachusett (2006'), which was once owned by his father's
father, the first minister of the village of Princeton.

"Before noon we had reached the highlands overlooking the valley of Lancaster (affording the first fair and
open prospect into the west), and there, on the top of a
hill, in the shade of some oaks, near to where a spring
bubbled out from a leaden pipe, we rested during the
heat of the day, reading Virgil and enjoying the scenery. It
was such a place as one feels to be on the outside of the
earth; for from it we could, in some measure, see the form
and structure of the globe." Howarth suggests that this is
Wataquadock Hill, though this entire escarpment affords
views west over Nashua Valley to Wachusett.

"The descent into the valley on the Nashua side is by

far the most sudden; and a couple of miles brought us to the southern branch of the Nashua, a shallow but rapid stream, flowing between high and gravelly banks....

"As we went on our way late in the afternoon, we refreshed ourselves by bathing our feet in every rill that crossed the road, and anon, as we were able to walk in the shadows of the hills, recovered our morning elasticity. Passing through Sterling, we reached the banks of the Stillwater, in the western part of the town, at evening, where is a small village collected.... This village had, as yet, no post-office, nor any settled name."

Early the next morning, they left their inn in west Sterling and walked 4 miles to the base of Wachusett, following for part of the way the course of the Stillwater River.

"As we gathered the raspberries, which grew abundantly by the roadside, we fancied that that action was consistent with a lofty prudence; as if the traveler who ascends into a mountainous region should fortify himself by eating of such light ambrosial fruits as grow there, and drinking of the springs which gush out from the mountain-sides, as he gradually inhales the subtler and purer atmosphere of those elevated places, thus propitiating the mountain gods by a sacrifice of their own fruits. The gross products of the plains and valleys are for such as dwell therein; but it seemed to us that the juices of this berry had relation to the thin air of the mountain-tops.

"In due time we began to ascend the mountain, passing, first, through a grand sugar maple wood, which bore the marks of the auger, then a denser forest, which gradually became dwarfed, till there were no trees whatever. We at length pitched our tent on the summit. It is but nineteen hundred feet above the village of Princeton, and three thousand above the level of the sea...." Actually, the summit is only 831' above the village.

"The summit consists of a few acres, destitute of trees, covered with bare rocks, interspersed with blueberry bushes, raspberries, gooseberries, strawberries, moss, and a fine, wiry grass. The common yellow lily and dwarf cornel grow abundantly in the crevices of the rocks. This clear space, which is gently rounded, is bounded a few feet lower by a thick shrubbery of oaks, with maples, aspens, beeches, cherries, and occasionally a mountain-ash intermingled, among which we found the bright blue berries of the Solomon's-seal, and the fruit of the pyrola. From the foundation of a wooden observatory, which was formerly erected on the highest point, forming a rude, hollow structure of stone, a dozen feet in diameter, and five or six in height, we could see Monadnock, in simple grandeur, in the northwest, rising nearly a thousand feet higher....

"We read Virgil and Wordsworth in our tent, with new pleasure there, while waiting for a clearer atmosphere....

"Who knows but this hill may one day be a Helvellyn, or even a Parnassus, and the Muses haunt here, and other Homers frequent the neighboring plains?...

"The blueberries which the mountain afforded, added to the milk we had brought, made our frugal supper, while for entertainment the even-song of the wood thrush rang along the ridge. Our eyes rested on no painted ceiling nor carpeted hall, but on skies of Nature's painting, and hills and forests of her embroidery. Before sunset, we rambled along the ridge to the north, while a hawk soared still above us. It was a place where gods might wander, so solemn and solitary, and removed from all contagion with the plain. As the evening came on, the haze was condensed in vapor, and the landscape became more distinctly visible, and numerous sheets of water were brought to light....

"There was, by chance, a fire blazing on Monadnock that night, which lighted up the whole western horizon, and, by making us aware of a community of mountains, made our position seem less solitary. But at length the wind drove us to the shelter of our tent, and we closed its door for the night, and fell asleep....

"The morning twilight began as soon as the moon had set, and we arose and kindled our fire, whose blaze might have been seen for thirty miles around. As the daylight increased, it was remarkable how rapidly the wind went down. There was no dew on the summit, but coldness supplied its place. When the dawn had reached its prime, we enjoyed the view of a distinct horizon line, and could fancy ourselves at sea, and the distant hills the waves in the horizon, as seen from the deck of a vessel. The cherry-birds [cedar waxwings] flitted around us, the nuthatch and flicker were heard among the bushes, the titmouse perched within a few feet, and the song of the wood thrush again rang along the ridge. At length we saw the sun rise up out of the sea, and shine on Massachusetts.... There was little of the sublimity and grandeur which belong to mountain scenery, but an immense landscape to ponder on a summer's day....

"We could at length realize the place mountains occupy on the land, and how they come into the general scheme of the universe. When first we climb their summits and observe their lesser irregularities, we do not give credit to the comprehensive intelligence which shaped them; but when afterward we behold their outlines in the horizon, we confess that the hand which moulded their opposite slopes, making one to balance the other, worked round a deep centre, and was privy to the plan of the universe. . . .

"At noon we descended the mountain. . . . Passing swiftly through Stillwater and Sterling, as with a downward impetus, we found ourselves almost at home again in the green meadows of Lancaster, so like our own Concord, for both are watered by two streams, which unite near their centres, and have many other features in common. There is an unexpected refinement about this scenery; level prairies of great extent, interspersed with elms and hop-fields and groves of trees, give it almost a classic appearance."

At sunset they arrived in the village of Still River, from where "the prospect is beautiful, and the grandeur of the mountain outlines unsurpassed. There was such a repose and quiet here at this hour, as if the very hillsides were enjoying the scene."

Just north of Still River, on Prospect Hill, Bronson Alcott and Charles Lane established Fruitlands in 1843. Though their utopian community lasted only seven months, it is preserved as a historic site and open to the public. Two months after Thoreau, 27–28 September 1842, Emerson and Hawthorne walked from Concord to Harvard for the night and continued on the next day three more miles to visit the Shakers. Where Thoreau and Fuller stayed in Harvard is not disclosed; Howarth suspects with the Shakers. The town of Harvard offers generous conservation lands, historic homes, views of

Monadnock, and the poems of resident Elizabeth R. Cooper in *The Harvard Post*.

Early the next morning, the travelers parted ways, Fuller to Groton, Thoreau to Concord. "And now that we have returned to the desultory life of the plain, let us endeavor to import a little of that mountain grandeur into it. We will remember within what walls we lie, and understand that this level life too has its summit, and why from the mountain-top the deepest valleys have a tinge of blue; that there is elevation in every hour, as no part of the earth is so low that the heavens may not be seen from, and we have only to stand on the summit of our hour to command an uninterrupted horizon."

TEXT: "A Walk to Wachusett," *The Boston Miscellany*,
January 1843, 31–36.

19–20 OCTOBER 1854

Thoreau returned to Wachusett on 19 October 1854, this time with Thomas Cholmondeley of England and H. G. O. Blake of Worcester. They went by train to Westminster and from there on foot, 4 miles to the Foster house and 2 miles more to the summit by road. Possibly the house belonged to Josiah and Julia Foster of Westminster, though the location of their extant home, at 37 Spruce Road, does not match Thoreau's mileage.

"The country above Littleton (plowed ground) more or less sugared with snow, the first I have seen. We find a little on the mountain-top. The prevailing tree on this mountain, top and all, is apparently the red oak, which toward and on the top is very low and spreading. Other trees and shrubs which I remember on the top are beech, *Populus tremuliformis* [quaking aspen, *P. tremuloides*], mountain-ash (looking somewhat like sumach), witch-hazel, white and yellow birch, white pine, black spruce, etc., etc. Most of the deciduous woods look as if dead. On the sides, beside red oak, are rock maple, yellow birch, lever-wood [eastern hop-hornbeam, *Ostrya virginica*], beech, chestnut, shagbark, hemlock, striped maple, witch-hazel, etc., etc.

"With a glass you can see vessels in Boston Harbor from the summit, just north of the Waltham Hills." They spent the night at the Foster house.

"*Oct.* 20. Saw the sun rise from the mountain-top. This is the time to look westward. All the villages, steeples, and houses on that side were revealed; but on the east all the landscape was a misty and gilded obscurity. It was worth the while to see westward the countless hills and fields all apparently flat, now white with frost. A little white fog marked the site of many a lake and the course of the Nashua, and in the east horizon the great pond had its own fog mark in a long, low bank of cloud.

"Soon after sunrise I saw the pyramidal shadow of the mountain reaching quite across the State, its apex resting on the Green or Hoosac Mountains, appearing as a deep-blue section of a cone there. It rapidly contracted, and its apex approached the mountain itself, and when about three miles distant the whole conical shadow was very distinct."

TEXT: *The Journal of Henry D. Thoreau,* VII:64–66.

Thoreau often observed Wachusett from Concord. "Wachuset [*sic*] free of clouds has a fine purplish tinge

—as if the juice of grapes had been squeezed over it—
darkening into blue" (*Journal,* 27 July 1852). He even
sketched it. "Wachusett from Fair Haven Hill looks like
this:—

the dotted line being the top of the surrounding forest"
(*Journal,*2 August 1852).

And, in another season, he produced this metaphor:
"It is a true winter sunset, almost cloudless, clear, cold
indigo-y along the horizon....A rosy tint suffuses the
eastern horizon. The outline of the mountains is wonderfully distinct and hard, and they are a dark blue and
very near. Wachusett looks like a right whale over our
bow, plowing the continent, with his flukes well down.
He has a vicious look, as if he had a harpoon in him"
(*Journal,* 27 December 1853).

⟩

You can take the train from Concord to Fitchburg, where
Thoreau also went to lecture in 1857. In 1859, Emerson
took a carriage from the station to the mountain, where
he sprained his foot. Walk west 5 miles from Fitchburg to
Westminster (which Thoreau did on 5 September 1856).
Westminster Hill Road avoids 2A traffic. From Westminster station, formerly north of 2A at the Depot Road in-
tersection, walk south 6.3 miles on Narrows, Stone Hill,
East, Gatehouse, Mile Hill, Park, and Mountain roads, to
Gregory Road on left, almost a mile south of the visitor
center. Opposite is Mountain House Trail, which rises a

mile to the crest, along a course close to the old Coast Survey Road, which Thoreau and Emerson used.

About a mile south on Mountain Road is Little Wachusett Mountain (the trail is off Westminster Road), which Emerson climbed on 1 April 1845 and "found a view which made my heart beat" (*Letters*, 8:18). Emerson's Wachusett excursions inspired poetry. From Little Wachusett it is 2 miles south and west to Wachusett Meadow Audubon Sanctuary, on Goodnow Road in Princeton. I prefer going from here over delightful Brown Hill to Dickens and Harrington trails north to Wachusett. This mountain also begs being walked around (7.5 miles). My friend Bill Brace makes a loop by bicycle from his Concord home to the top of Wachusett and back within a day.

Greylock

♦

? JULY–1 AUGUST 1844

Thoreau's thirty-seventh birthday was 12 July. Margaret Fuller wrote on 13 July of "a very pleasant visit from Henry" in Concord. Probably he departed on the fourteenth.

Greylock is part of an ancient pathway connecting the Atlantic Ocean and the Hudson River which European settlers called the Mohawk Trail. The trail followed the Deerfield River west from the Connecticut River, over the Hoosac Mountains, down the Hoosic River to the

Hudson at Schaghticoke, New York. Having made his way from Monadnock, probably by way of the Ashuelot River valley, to the Connecticut River at Hinsdale, New Hampshire, Thoreau approached the Berkshires over this route. Now you can walk some of it along a new recreational trail.

"I had come over the hills on foot and alone in serene summer days, plucking the raspberries by the way-side, and occasionally buying a loaf of bread at a farmer's house, with a knapsack on my back, which held a few traveller's books and a change of clothing, and a staff in my hand. I had that morning looked down from the Hoosack Mountain, where the road crosses it, on the village of North Adams in the valley, three miles away under my feet, showing how uneven the earth may sometimes be, and making it seem an accident that it should ever be level and convenient for the feet of man."

This is a spectacular view, from West Summit (2018') of the Hoosac Range, 4.5 miles east of North Adams, where Thoreau stopped for rice, sugar, and a tin cup.

"I began in the afternoon to ascend the mountain whose summit is three thousand six hundred feet above the level of the sea, and was seven or eight miles distant by the path." Greylock (a.k.a. Saddleback) is actually 109' shorter, though its height was not established until a century after Thoreau. His mileage estimate is accurate.

"My route lay up a long and spacious valley called the Bellows, because the winds rush up or down it with violence in storms, sloping up to the very clouds between the principal range and a lower mountain. There were a few farms scattered along at different elevations, each commanding a fine prospect of the mountains to the north, and a stream ran down the middle of the valley, on which near the head there was a mill. It seemed a road for the pilgrim to enter upon who would climb to the gates of heaven. Now I crossed a hay-field, and now over the brook on a slight bridge, still gradually ascending all the while, with a sort of awe, and filled with indefinite expectations as to what kind of inhabitants and what kind of nature I should come to at last."

You can follow the Bellows Pipe Trail from the Notch Road to the Appalachian Trail, 3.7 miles.

"I at length reached the last house but one, where the path to the summit diverged to the right, while the summit itself rose directly in front. But I determined to follow up the valley to its head, and then find my own route up the steep, as the shorter and more adventurous way."

A comely woman of that house told him that Williams College students ascended the mountain "almost every pleasant day," but never by Thoreau's way.

"As I passed the last house, a man called out to know what I had to sell, for seeing my knapsack, he thought that I might be a pedler, who was taking this unusual route over the ridge of the valley into South Adams. He told me that it was still four or five miles to the summit by the path which I had left, though not more than two in a straight line from where I was, but that nobody ever went this way; there was no path, and I should find it as steep as the roof of a house. But I knew that I was more used to woods and mountains than he, and went along through his cow-yard, while he, looking at the sun, shouted after me that I should not get to the top that night." Thoreau left the path and went to the top by compass.

"The ascent was by no means difficult or unpleasant,

and occupied much less time than it would have taken to follow the path. Even country people, I have observed, magnify the difficulty of travelling in the forest, especially among mountains. They seem to lack their usual common sense in this. I have climbed several higher mountains without guide or path, and have found, as might be expected, that it takes only more time and patience commonly than to travel the smoothest highway. It is very rare that you meet with obstacles in this world, which the humblest man has not faculties to surmount."

Thoreau's "several higher mountains," at the time this was written, were Washington and Katahdin. He reached Greylock's summit at sunset, a long day — about 16 miles from Luke Rice's farm in Florida, from where he departed before anyone was awake. He was "half way over the mountain with the sun." This is Whitcomb Summit (2173') of the Hoosac Range.

He camped next to the observatory, "a building of considerable size, erected by the students of Williamstown College, whose buildings might be seen by daylight gleaming far down in the valley. It would be no small advantage if every college were thus located at the base of a mountain, as good at least as one well-endowed professorship. It were as well to be educated in the shadow of a mountain as in more classical shades. Some will remember, no doubt, not only that they went to the college, but that they went to the mountain. Every visit to its summit would, as it were, generalize the particular information gained below, and subject it to more catholic tests."

When Thoreau saw Williams College it was still in session, with 155 students, mostly from New York and New England (none from Concord), and seven professors and two tutors. Mark Hopkins served as both professor and president; his brother, Albert, professor of natural philosophy and astronomy, started the Horticultural and Landscape Gardening Association to encourage health, campus beautification, and divine communion, and founded New England's first Alpine Club in 1863. Prayers were at sunrise and evening, chapel at noon. Mountain Day was celebrated in spring. Tuition was ten dollars a term. The tower of Thoreau's acquaintance, erected in 1841 and pictured in *Most Excellent Majesty,* is no longer, but Mountain Day still happens.

Hopkins was still president twenty-one years later, when Emerson came to give a lecture in November 1865. His enthusiastic reception prompted six talks. One morning, with Charles J. Woodbury and other students, he walked to Greylock. Apparently they did not ascend it — Greylock, "a serious mountain" to Emerson, reminded him of Wordsworth's poem *The Excursion* and of the hills of Westmoreland. "The landscape is good for the eyes every hour of the day," he wrote his daughter Ellen, "with its frosty morning mountains, its noon purple glooms, & serious invitations to the feet" (*Letters,* 5:434).

Thoreau "lay down on a board...not having any blanket to cover me.... But as it grew colder towards midnight, I at length encased myself completely in boards, managing even to put a board on top of me, with a large stone on it, to keep it down, and so slept comfortably....

"I was up early and perched upon the top of this tower to see the daybreak.... As the light increased I discovered around me an ocean of mist, which by chance reached up exactly to the base of the tower, and shut out every vestige of the earth . . . it revealed to me more clearly the new world into which I had risen in the night, the new terra-firma perchance of my future life. There was not a crevice left through which the trivial places we name Massachusetts, or Vermont, or New York, could be seen, while I still inhaled the clear atmosphere of a July morning, — if it were July there. All around beneath me was spread for a hundred miles on every side, as far as the eye could reach, an undulating country of clouds, answering in the varied swell of its surface to the terrestrial world it veiled. It was such a country as we might see in dreams, with all the delights of paradise. . . . It was a favor for which to be forever silent to be shown this vision."

After this epiphany Thoreau descended, steering south by compass for Pontoosuc Lake, now partly within the city of Pittsfield. You can approximate his route by taking the Appalachian Trail to Jones Nose Trail, to Rockwell Road, to Quarry Road, to Pittsfield via North Main Street and Route 7, passing Pontoosuc Lake, or stay with the Appalachian Trail to Cheshire and Dalton, which is just east of Pittsfield.

Later that morning, having walked more than 15 miles, he met Ellery Channing at Pittsfield's railroad station, on the west side of North Street. This depot burned in November 1854. Thoreau missed the address by Oliver Wendell Holmes at the Berkshire Jubilee by a month.

Three miles south of the station is Arrowhead, the Pittsfield home of Herman Melville from 1850 to 1863. Thoreau passed it on the train en route to Chatham and Hudson, New York. Seven years later, in August 1851, Melville and his party spent a night atop Greylock in the observatory, bundled in quilts and buffalo robes. During this convivial venture Melville read Thoreau's Greylock excursion in *A Week*.

TEXT: "Tuesday" chapter of
A Week on the Concord and Merrimack Rivers.

Catskills

♦

LATE JULY 1844

From Greylock, Thoreau saw the Catskills in New York, some sixty miles to the southwest, as "new and yet higher mountains." Though the Catskills do indeed rise above the Berkshires, culminating in Slide Mountain (4180'), Thoreau attained lower elevations in them. To get to the Catskills from Pittsfield, he went with Ellery Channing by rail to Hudson, New York, where they took an overnight steamboat up the Hudson to Albany (Thoreau, en route to Minnesota, stayed at the Delevan House in Albany on 13 May 1861), later going downriver to the west-bank village of Catskill, opposite Hudson.

This was a propitious moment. When Thoreau came through the village of Catskill, two artists resided there: Thomas Cole, who had lived there since 1836 and had painted the Catskills since 1825, and the eighteen-year-old Frederic Edwin Church, fresh from Hartford in early June to study with Cole for two years. Cole and his wife and family rented Cedar Grove, at 218 Spring Street, (which still stands), the home of John Alexander Thomson, the uncle of Cole's wife, Maria Bartow; Church also boarded on Thomson's farm. Apparently Cole and Church had just returned from five days at the Catskill Mountain House, where Thoreau was headed.

During this time, the teacher exposed his student to his favorite sketching places. Red Hill, across the Hudson River from Catskill, Church found irresistible. In 1860 he purchased 126 acres on it and later acquired higher ground on which he constructed a house, Olana, where he lived until his death. As a state historic site, Olana is open to the public from April to October.

From Catskill, Thoreau went west for 12 miles via Mountain House Road, up the eastern escarpment (we don't know how he got there, perhaps by stage, perhaps on foot) to Kaaterskill Falls, where later, in his Walden Journal (II:155), he described his residence:

"Yesterday I came here to live. My house makes me think of some mountain houses I have seen, which seemed to have a fresher auroral atmosphere about them as I fancy of the halls of Olympus. I lodged at the house of a saw-miller last summer, on the Caatskills mountains, high up as Pine orchard in the blue-berry & raspberry region, where the quiet and cleanliness & coolness seemed to be all one, which had this ambrosial character. He was the miller of the Kaaterskill Falls. They were a clean & wholesome family inside and out — like their house. The latter was not plastered — only lathed and the inner doors were not hung. The house seemed high placed, airy, and perfumed, fit to entertain a travelling God. It was so high indeed that all the music, the broken strains, the waifs & accompaniments of tunes, that swept over the ridge of the Caatskills, passed through its aisles."

The historian Alf Evers identifies Thoreau's hosts as Ira and Mary Scribner. Ira operated a sawmill on Lake

Creek, which, less than a mile from South and North lakes, flows over Kaaterskill Falls. Mary ran their home, Glen Mary Cottage, as an inn. It was under construction when Thoreau arrived, so he must have been one of their first, if not the first, lodger. "Mary Scribner set a good table," according to Justine Hommel, "kept a smile, clean house and charged very reasonable rates." It was near Pine Orchard (2200'), where the Catskill Mountain House (1823–1963) was located and from which walking trails led to North (3180') and South (2460') mountains and intervale lakes. This is now a state-owned campground. Their inn, which expanded over the years, is gone. North of North Lake, Mary's Glen Trail celebrates Mary Scribner.

We can imagine Thoreau walking the 1.5-mile trail from Kaaterskill Falls to the Catskill Mountain House, taking in the sumptuous view of the Hudson River valley, ascending North and South mountains, bathing in the lakes.

Southern Berkshires

Thoreau and Channing returned via Bash Bish Falls in southwestern Massachusetts, again over the Berkshires and through the villages of Mount Washington and Chester. Likely they passed, and at least viewed, Monument Mountain, in the towns of Stockbridge and Great Barrington, the meeting place of Melville and Hawthorne on 5 August 1850. At Chester they took the railroad east, arriving in Concord on August 1.

Two weeks after his return, while declining Isaac

Hecker's invitation to tour Europe afoot, Thoreau ruminated about travel and the pull of place and spirit:

"I have but just returned from a pedestrian excursion, some what similar to that you propose, *parvis componere magna* ["to compare great things with small," Virgil, *Eclogues*], to the Catskill mountains, over the principal mountains of this state, subsisting mainly on bread and berries, and slumbering on the mountain tops. As usually happens, I now feel a slight sense of dissipation. Still I am strongly tempted by your proposal and experience a decided schism between my outward and inward tendencies. Your method of travelling especially — to *live* along the road — citizens of the world, without haste or petty plans — I have often proposed this to my dreams, and still do — But the fact is, I cannot so decidedly postpone exploring the *Farther Indies,* which are to be reached you know by other routs and other methods of travel. I mean that I constantly return from every external enterprise with disgust to fresh faith in a kind of Brahminical Artesian, Inner Temple, life. All my experience, as yours probably, proves only this reality" (*C,* 155–56).

Katahdin

♦

31 August–11 September 1846

In the mid-nineteenth century, getting to Katahdin was as much an ordeal as climbing the mountain itself. While living at Walden, Thoreau traveled to Katahdin and back within a fortnight. A three-dollar steamship passage brought him from Boston to Bangor, Maine. Then he went with his cousin George Thatcher by horse and carriage north along the Penobscot River to Mattawamkeag. From there it was a seven-mile side trip to Molunkus, the farthest east Thoreau had been, where a public house was the only structure in town, "but sometimes even this is filled with travellers." The next day they walked "an obscure trail up the northern bank of the Penobscot" into "a wholly uninhabited wilderness, stretching to Canada" until they reached George McCauslin's farm. There they awaited an Indian guide, Louis Neptune, who disappointed them. McCauslin went along instead, with "fifteen pounds of hard bread, ten pounds of 'clear' pork, and a little tea" in his pack. By bateau, their party of six went up the West Branch, commingled with beautiful lakes. "The country is an archipelago of lakes, — the lake-country of New England."

Millinocket and Ambajejus lakes almost touch. Frederic Edwin Church, who first climbed and sketched Katahdin in August 1852 — producing that September *Mount Katahdin from Katahdin Lake,* from which derived his famous *Mount Ktaadn* — summered on Millinocket Lake.

For the last 9 miles of their voyage, "we rowed across several small lakes, poled up numerous rapids and thoroughfares, and carried over four portages."

At Abol Stream they left their bateau and proceeded on foot. "We determined to steer directly for the base of the highest peak, leaving a large slide [Abol] ... on our left. ... Ktaadn presented a different aspect from any mountain I have seen, there being a greater proportion of naked rock, rising abruptly from the forest; and we looked up at this blue barrier as if it were some fragment of a wall which anciently bounded the earth in that direction. Setting the compass for a north-east course, which was the bearing of the southern base of the highest peak, we were soon buried in the woods."

Baxter Peak (5267') hides behind South Peak (5240') from this vantage point.

"The tracks of moose, more or less recent, to speak literally covered every square rod on the sides of the mountain.... Sometimes we found ourselves travelling in faint paths, which they had made ... everywhere the twigs had been browsed by them, clipt as smoothly as if by a knife. The bark of trees was stript up by them to the height of eight or nine feet, in long narrow strips, an inch wide, still showing the distinct marks of their teeth. We expected nothing less than to meet a herd of them every moment....

"By the side of a cool mountain rill, amid the woods, where the water began to partake of the purity and transparency of the air, we stopped to cook some of our fishes, which we had brought thus far in order to save our hard bread and pork, in the use of which we had put ourselves on short allowance. We soon had a fire blazing....

"The wood was chiefly yellow birch, spruce, fir, mountain-ash, or round-wood, as the Maine people call it, and moose-wood [striped maple, *Acer pensylvanicum*]. It was the worst kind of travelling.... The cornel, or bunch-berries, were very abundant, as well as Solomon's seal and moose-berries [hobblebush, *Viburnum alnifolium*]. Blue-berries were distributed along our whole route; and in one place the bushes were drooping with the weight of the fruit, still as fresh as ever. ... Such patches afforded a grateful repast, and served to bait the tired party forward."

From their camp, made at four that afternoon near the treeline (about 3800') in a ravine west of Rum Mountain, Thoreau scaled Katahdin. "Following up the course of the torrent which occupied this [ravine] — and I mean to lay some emphasis on this word *up* — pulling myself up by the side of perpendicular falls of twenty or thirty feet, by the roots of firs and birches, and then, perhaps, walking a level rod or two in the thin stream, for it took up the whole road, ascending by huge steps, as it were, a giant's stairway, down which a river flowed, I had soon cleared the trees, and paused on the successive shelves, to look back over the country.... Having slumped, scrambled, rolled, bounced, and walked, by turns, over this scraggy country, I arrived upon a side-hill, or rather side-mountain, where rocks, gray, silent rocks, were the flocks and herds that pastured, chewing a rocky cud at sunset. They looked at me with hard gray eyes, without a bleat or a low. This brought me to the skirt of a cloud, and bounded my walk that night."

Thoreau then returned to his companions at camp. "As here was no cedar, we made our bed of coarser feathered spruce; but at any rate the feathers were plucked from the live tree. It was, perhaps, even a more grand and desolate place for a night's lodging than the summit would have been, being in the neighborhood of those wild trees, and of the torrent....

"In the morning [8 September], after whetting our appetite on some raw pork, a wafer of hard bread, and a dipper of condensed cloud or water-spout, we all together began to make our way up the falls, which I have described; this time choosing the right hand, or highest peak, which was not the one I had approached before. But soon my companions were lost to my sight behind the mountain ridge in my rear, which still seemed ever retreating before me, and I climbed alone over huge rocks, loosely poised, a mile or more, still edging toward the clouds—for though the day was clear elsewhere, the summit was concealed by mist. The mountain seemed a vast aggregation of loose rocks, as if sometime it had rained rocks, and they lay as they fell on the mountain sides, nowhere fairly at rest, but leaning on each other, all rocking-stones, with cavities between, but scarcely any soil or smoother shelf. They were the raw materials of a planet dropped from an unseen quarry, which the vast chemistry of nature would anon work up, or work down, into the smiling and verdant plains and valleys of earth. This was an undone extremity of the globe."

On the "summit of the ridge"—I place him between South and Baxter peaks—he was enveloped in clouds. "Now the wind would blow me out a yard of clear sunlight, wherein I stood; then a gray, dawning light was all it could accomplish, the cloud-line ever rising and falling with the wind's intensity. Sometimes it seemed as if the summit would be cleared in a few moments and smile in sunshine: but what was gained on one side was lost on another. It was like sitting in a chimney and waiting for the smoke to blow away. It was, in fact, a cloud-factory,— these were the cloud-works, and the wind turned them off done from the cool, bare rocks. Occasionally, when the windy columns broke in to me, I caught sight of a dark, damp crag to the right or left; the mist driving ceaselessly between it and me. It reminded me of the creations of the old epic and dramatic poets, of Atlas, Vulcan, the Cyclops, and Prometheus. Such was Caucasus and the rock where Prometheus was bound. Æschylus had no doubt visited such scenery as this. It was vast, Titanic, and such as man never inhabits. Some part of the beholder, even some vital part, seems to escape through the loose grating of his ribs as he ascends. He is more lone than you can imagine....

"The tops of mountains are among the unfinished parts of the globe, whither it is a slight insult to the gods to climb and pry into their secrets, and try their effect on our humanity. Only daring and insolent men, perchance, go there. Simple races, as savages, do not climb mountains — their tops are sacred and mysterious tracts never visited by them. Pomola is always angry with those who climb to the summit of Ktaadn." To the Penobscot, Pomola is a spirit of the mountain.

Giving up hope of any clearing, Thoreau descended. "I found my companions where I had left them, on the side of the peak, gathering the mountain cranberries, which filled every crevice between the rocks, together with blue berries, which had a spicier flavor the higher up they grew, but were not the less agreeable to our palates. When the country is settled and roads are made, these cranberries will perhaps become an article of commerce. From this elevation, just on the skirts of the clouds, we could overlook the country west and south for a hundred miles. There it was, the State of Maine, which we had seen on the map, but not much like that. Immeasurable forest for the sun to shine on, that eastern *stuff* we hear of in Massachusetts. No clearing, no house. It did not look as if a solitary traveller had cut so much as a walking-stick there.

"Countless lakes, — Moosehead in the southwest, forty miles long by ten wide, like a gleaming silver platter at the end of the table; Chesuncook, eighteen long by three wide, without an island; Millinocket, on the south, with its hundred islands; and a hundred others without a name; and mountains also, whose names, for the most part, are known only to the Indians. The forest looked like a firm grass sward, and the effect of these lakes in its midst has been well compared by one who has since visited this same spot, to that of a 'mirror broken into a thousand fragments, and wildly scattered over the grass, reflecting the full blaze of the sun' " (J. K. Laski).

Their return was along Abol Stream, "continually crossing and recrossing it, leaping from rock to rock, and jumping with the stream down falls of seven or eight feet, or sometimes sliding down on our backs in a thin sheet of water.... We travelled thus very rapidly with a downward impetus, and grew remarkably expert at leaping from rock to rock, for leap we must, and leap we did, whether there was any rock at the right distance or not. ... The cool air above, and the continual bathing of our bodies in mountain water, alternate foot, sitz, douche, and plunge baths, made this walk exceedingly refreshing, and we had travelled only a mile or two after leaving the torrent, before every thread of our clothes was as dry as usual, owing perhaps to a peculiar quality in the atmosphere....

"Perhaps I most fully realized that this was primeval, untamed, and forever untameable *Nature*...while coming down this part of the mountain.... It is difficult to conceive of a region uninhabited by man. We habitually presume his presence and influence everywhere. And yet we have not seen pure Nature, unless we have seen her thus vast, and drear, and inhuman, though in the midst of cities. Nature was here something savage and awful, though beautiful. I looked with awe at the ground I trod on, to see what the Powers had made there, the form and fashion and material of their work. This was that Earth of which we have heard, made out of Chaos and Old Night. Here was no man's garden, but the unhandselled globe. It was not lawn, nor pasture, nor mead, nor woodland, nor lea, nor arable, nor waste-land. It was the fresh and natural surface of the planet Earth.... Think of our life in nature,—daily to be shown matter, to come in contact with it,—rocks, trees, wind on our cheeks! the *solid* earth! the *actual* world! the *common sense! Contact! Contact! Who* are we? *where* are we?"

By two, they reached their bateau at the mouth of Abol Stream. There they had dinner, and at four started down the Penobscot's West Branch. "Though we glided so swiftly and often smoothly down, where it had cost us no slight effort to get up, our present voyage was attended with far more danger: for if we once fairly struck one of the thousand rocks by which we were surrounded, the boat would be swamped in an instant." They arrived without mishap in Bangor three days later. Soon Thoreau "was steaming his way to Massachusetts."

Thoreau ended with what Joseph Moldenhauer calls "a philosophical coda," three and a quarter printed pages that consider wilderness: "What is most striking in the Maine wilderness is, the continuousness of the forest.... Except the few burnt lands, the narrow intervals on the rivers, the bare tops of the high mountains, and the lakes and streams, the forest is uninterrupted. It is ever more grim and wild than you had anticipated, a damp and intricate wilderness....

"I am reminded by my journey how exceedingly new this country still is.... America is still unsettled and unexplored...we live only on the shores of a continent even yet, and hardly know where the rivers come from which float our navy....

"We have advanced by leaps to the Pacific, and left many a lesser Oregon and California unexplored behind us. Though the railroad and the telegraph have been established on the shores of Maine, the Indian still looks out from her interior mountains over all these to the sea."

TEXT: "Ktaadn" chapter of *The Maine Woods.*

Kineo

♦

20 JULY–8 AUGUST 1857

Thoreau returned to Maine to lecture in Portland on 21 March 1849 and 15 January 1851, and to see its interior for fifteen September days in 1853, but he did not climb any mountains on these trips. Even the latter journey, which brought him by Kineo on a steamboat passage up and down the length of Moosehead Lake, permitted only distant views. "The scenery is not merely wild, but varied and interesting; mountains were seen further or nearer on all sides but the north-west, their summits now lost in the clouds, but Mount Kineo is the principal feature of the lake, and more exclusively belongs to it....

"Mount Kineo, at which the boat touched, is a peninsula with a narrow neck, about mid-way the lake on the east side. The celebrated precipice is on the east or land side of this, and is so high [1806'] and perpendicular that you can jump from the top many hundred feet into the water which makes up behind the point."

His last Maine journey, which brought him again to Moosehead Lake, this time in a birch-bark canoe with a Penobscot Indian guide, Joe Polis, and the Concord attorney Edward Hoar, did involve an overnight stay at Kineo and, at least for the tourists, a view from its summit. At the start, "Squaw Mountain rose darkly on our left... and what the Indian called Spencer Bay Mountain, on the east, and already we saw Mount Kineo before us in the north." This alpine encirclement graces New England's largest lake. You can paddle after Thoreau on Moosehead Lake from Greenville to Kineo, 17.5 miles; or make the half-mile trip east across Moosehead to Kineo from Rockwood, 19 miles north of Greenville.

"While we were crossing this bay, where Mount Kineo rose dark before us, within two or three miles, the Indian repeated the tradition respecting this mountain's having anciently been a cow moose, — how a mighty Indian hunter, whose name I forget, succeeded in killing this queen of the moose tribe with great difficulty, while her calf was killed somewhere among the islands in Penobscot Bay, and, to his eyes, this mountain had still the form of the moose in a reclining posture, its precipitous side presenting the outline of her head....

"We approached the land again through pretty rough

water, and then steered directly across the lake, at its narrowest part, to the eastern side, and were soon partly under the lee of the mountain, about a mile north of the Kineo House, having paddled about twenty miles. It was now about noon.

"We designed to stop there that afternoon and night, and spent half an hour looking along the shore northward for a suitable place to camp.... At length, half a mile further north, by going half a dozen rods into the dense spruce and fir wood on the side of the mountain, almost as dark as a cellar, we found a place sufficiently clear and level to lie down on, after cutting away a few bushes....

"After dinner we returned southward along the shore, in the canoe, on account of the difficulty of climbing over the rocks and fallen trees, and began to ascend the mountain along the edge of the precipice. But a smart shower coming up just then, the Indian crept under his canoe, while we, being protected by our rubber coats, proceeded to botanize. So we sent him back to the camp for shelter, agreeing that he should come there for us with his canoe toward night. It had rained a little in the forenoon, and we trusted that this would be the clearing-up shower, which it proved; but our feet and legs were thoroughly wet by the bushes. The clouds breaking away a little, we had a glorious wild view, as we ascended, of the broad lake with its fluctuating surface and numerous forest-clad islands, extending beyond our sight both north and south, and the boundless forest undulating away from its shores on every side, as densely packed as a rye-field, and enveloping nameless mountains in succession; but above all, looking westward over a large island was visible a very distant part of the lake, though we did not then suspect it to be Moosehead, — at first a mere broken white line seen through the tops of the island trees, like haycaps, but spreading to a lake when we got higher. Beyond this we saw what appears to be called Bald Mountain on the map, some twenty-five miles distant, near the sources of the Penobscot. It was a perfect lake of the woods. But this was only a transient gleam, for the rain was not quite over.

"Looking southward, the heavens were completely overcast, the mountains capped with clouds, and the lake generally wore a dark and stormy appearance....

"If I wished to see a mountain or other scenery under the most favorable auspices, I would go to it in foul weather, so as to be there when it cleared up; we are then in the most suitable mood, and nature is more fresh and inspiring. There is no serenity so fair as that which is just established in a tearful eye."

Kineo was composed of hornstone, "the largest mass of this stone known in the world," according to Maine's first state geologist, Charles T. Jackson, Emerson's brother-in-law, whom Thoreau quoted. Hornstone, an obsolete term, is a rhyolite of which larger masses are now known — for example, The Traveler Mountain north of Katahdin. Thoreau had "found hundreds of arrowheads made of the same material. It is generally slate-colored, with white specks, becoming a uniform white where exposed to the light and air, and it breaks with a conchoidal fracture, producing a ragged cutting edge....

"From the summit of the precipice which forms the southern and eastern sides of this mountain peninsula, and is its most remarkable feature, being described as five or six hundred feet high [today 777'], we looked, and probably might have jumped down to the water, or to the seemingly dwarfish trees on the narrow neck of land which connects it with the main. It is a dangerous place to try the steadiness of your nerves...

"The plants which chiefly attracted our attention on this mountain were the mountain cinquefoil (*Potentilla tridentata*), abundant and in bloom still at the very base, by the water-side, though it is usually confined to the summits of mountains in our latitude; very beautiful harebells [*Campanula rotundifolia*] overhanging the precipice; bear-berry [*Arctosthaphylos uva-ursi*]; the Canada blueberry (*Vaccinium Canadense*) [velvetleaf blueberry, *Vaccinium myrtilloides*], similar to the *V. Pennsylvanicum* [early low blueberry, *V. angustifolium*], our earliest one, but entire leaved and with a downy stem and leaf; I have not seen it in Massachusetts; *Diervilla trifida* [bush honeysuckle, *D. lonicera*]; *Microstylis ophioglossoides* [green adder's-mouth, *Malaxis unifolia*], an orchidaceous plant new to us; wild holly (*Nemopanthes Canadensis*) [*N. mucronata*]; the great round-leaved orchis (*Platanthera orbiculata*) [*Habenaria macrophylla*], not long in bloom; *Spiranthes cernua* [nodding ladies'-tresses], at the top; bunch-berry, reddening as we ascended, green at the base of the mountain, red at the top; and the small fern, *Woodsia ilvensis* [rusty woodsia], growing in tufts, now in fruit. I have also received *Liparis liliifolia* [*sic*], or twayblade, from this spot. Having explored the wonders of the mountain, and the weather being now entirely cleared up, we commenced the descent. We met the Indian, puffing and panting, about one third of the way up, but thinking that he must be near the top, and saying that it took his breath away. I thought that superstition had something to do with his fatigue. Perhaps he believed that he was climbing over the back of a tremendous moose. He said that he had never ascended Kineo. On reaching the canoe we found that he had caught a lake trout weighing about three pounds, at the depth of twenty-five or thirty feet, while we were on the mountain."

Waking that night, Thoreau rejoiced in discovering phosphorescent wood, "which suggested to me that there was something to be seen if one had eyes. It made a believer of me more than before. I believed that the woods were not tenantless, but choke-full of honest spirits as good as myself any day,—not an empty chamber, in which chemistry was left to work alone, but an inhabited house,—and for a few moments I enjoyed fellowship with them."

TEXT: "Chesuncook" and "The Allegash and East Branch" chapters of *The Maine Woods*.

Uncanoonuc

♦

4–7 SEPTEMBER 1848

"Uncannunuc Mountain in Goffstown was visible from Amoskeag, five or six miles westward. It is the north-easternmost in the horizon, which we see from our native town, but seen from there is too ethereally blue to be the same which the like of us have ever climbed. Its name is

said to mean 'The Two Breasts,' there being two eminences some distance apart. The highest, which is about
fourteen hundred feet above the sea [North Peak, 1329'],
probably affords a more extensive view of the Merrimack
valley and the adjacent country than any other hill,
though it is somewhat obstructed by woods. Only a few
short reaches of the river are visible, but you can trace its
course far down stream by the sandy tracts on its banks."

Before taking this walking tour with Ellery Channing,
Thoreau had seen Uncanoonuc from Concord, from the
Merrimack River in September 1839, and possibly from
Wachusett. Later he recognized it from the Peterborough
Hills (South Pack) on 6 September 1852, undoubtedly
from Monadnock, and in passing to its west by carriage
on 18 July 1858.

Their itinerary went from Concord to Tyngsboro
(17.5 miles) and Dunstable to Moore's Falls (16 miles) of
the Merrimack River south of Manchester, New Hampshire (Thoreau locked through here in 1839). I presume
they camped near the falls, for 33.5 miles was a long day's
walk, even for the thirty-one-year-old Thoreau. The next
day they headed west overland for 9 miles (direct distance, or 12 miles via roads) to Uncanoonuc, from where
they saw Agamenticus Hill (691') in York, Maine; Joe
English Hill (1288'), 4 miles west in New Boston; Kearsarge (2937') beyond, now with a communications tower
for cellular phones; and north to Gunstock (2250') of the
Belknap Range, southwest of Lake Winnipesaukee, with
the White Mountains afar.

As North and South peaks are close to each other,
Thoreau may have crossed both. Mountain Road divides
them, and combined with Bog Brook Road provides an

opportunity for circumambulation of North Peak, despite the intrusion of many homes. Witch hazel blooms gloriously in October. A road ascends South Peak (1321') now, as well as a nature trail. The trail, a peaceful ramble from the beach at Mountain Base Pond, offers views of glacial erratics, which Thoreau may have noticed, and splendid hemlocks and beeches. The summit, a jumble of fenced transmission towers, does not invite contemplation; North Peak is better for that.

Descending, they walked 2 miles north to Goffstown, a small village bisected by Piscataquog River, and from there northeastward 10.5 miles to Hooksett on the Merrimack River, where they spent the night. The next morning they ascended the Pinnacle, "a small wooded hill which rises very abruptly to the height of about two hundred feet [484'] near the shore at Hooksett Falls. As Uncannunuc Mountain is perhaps the best point from which to view the valley of the Merrimack, so this hill affords the best view of the river itself. I have sat upon its summit [6 September 1848], a precipitous rock only a few rods long, in fairer weather, when the sun was setting and filling the river valley with a flood of light. You can see up and down the Merrimack several miles each way. The broad and straight river, full of light and life, with its sparkling and foaming falls, the islet which divides the stream, the village of Hooksett on the shore almost directly under your feet, so near that you can converse with its inhabitants or throw a stone into its yards, the woodland lake at its western base [Pinnacle Pond], and the mountains in the north and north-east, make a scene of rare beauty and completeness, which the traveller should take pains to behold."

In 1984, I stopped at George Robie's General Store in Hooksett to ask directions to the Pinnacle. "Ask him," Robie said, pointing to the hill's owner, who happened to be drinking coffee at one of the two tables. I introduced myself to Arthur Locke, who told me that his grandfather bought the Pinnacle in 1870, built a dance hall atop it, which could be reached by a road. "There was a sixty-foot tower on it from which you could see Portsmouth and Boston and supposedly the White Mountains. I never got to go up it, so I don't know, but that's what people said." He gave me directions and permission: "Take 3A north past the American Legion, next left onto Pine Street, then next left onto Ardon Road, which dead ends. Take the old road from there to the summit." At the turnabout at the end of Ardon Road, we found a path to Pinnacle Pond and along its east side, though Interstate 93 prevented any walk around. Our prospect affirmed Thoreau's.

POSTSCRIPT: Arthur Locke still owns the Pinnacle, which he plans to preserve forever. He continues to work at the library and has just finished a memoir of his experiences as a POW in Japan.

From Hooksett, Thoreau went south to Hampstead, lodging at Caleb Harriman's tavern on East Road—no longer extant—before returning to Concord the next day via Plaistow and Haverhill.

Text: "Wednesday" and "Thursday" chapters of
A Week on the Concord and Merrimack Rivers.

Wantastiquet

▶

5–12 SEPTEMBER 1856

"The village of Brattleboro is peculiar for the nearness of the primitive wood and the mountain... this everlasting mountain is forever lowering over the village, shortening the day and wearing a misty cap each morning. You look up to its top at a steep angle from the village streets. A great part belongs to the Insane Asylum. This town will be convicted of folly if they ever permit this mountain to be laid bare."

Wantastiquet (a.k.a. Chesterfield, 1351') is the prominent feature of Brattleboro, Vermont, though it is situated across the Connecticut River in Hinsdale and Chesterfield, New Hampshire. For fifteen years I have watched this mountain's daily changes. October colors its green face, beginning with the yellowing of white birches and culminating in the coppering of oaks. Illuminated by low, soft, clear afternoon autumn light and seen beyond yellow maples, brown tamaracks, black water, it is transcendental.

Wantastiquet is preserved forever wild. The Insane Asylum, which Thoreau walked past, founded in 1834 by the estate of Anna Hunt Marsh of Hinsdale, began buying parcels of the mountain in 1849 at ten dollars an acre. In 1943, the Brattleboro Retreat, the Asylum's successor, gave Wantastiquet to the state of New Hampshire for conservation.

Thoreau spent four nights in the village, staying at the

home of Ann and Addison Brown at 12 Chase Street—
the house still stands. Addison Brown (1799–1872) had
graduated from Harvard and its divinity school, married
Ann Elizabeth Wetherbee, and moved to Brattleboro,
where he was pastor of the Unitarian church, a superin-
tendent of schools, and for the last decade of his life the
editor of the *Vermont Phoenix*. Ann (1807–1906), an avid
reader who surely savored *Walden*, an expert botanist
with "a bright, strong, sunny nature, sweet and good to
the core" (Cabot, 396), had invited Thoreau to their
home (*Log*, 25 January 1855).

Thoreau walked daily here along the Connecticut River, West River, Coldwater Path, and Whetstone
Brook. The ramble beside the brook was de rigueur
for patients of Dr. Robert Wesselhoeft, who founded
the Brattleboro Water Cure in 1845. The next March,
Harriet Beecher Stowe was one of several hundred who
came that year to bathe, drink cold water, and walk the
new paths that Thoreau would take later. Also taking the
cure with Stowe was Caroline Sturgis, whom Emerson
stopped to see on his return to Concord from Middlebury College, where he delivered the commencement address in July 1845 (*Letters*, 15 July 1845).

Thoreau found this to be good botanical ground. He
identified many plants, pressed samples, ate raspberries.
His highest moment came when he discovered a seven-
foot bush, leatherwood, *Dirca palustris*, "the largest on
the low interval" beside Whetstone Brook, "its leaves
broad, like entire sassafras leaves; now beginning to turn
yellow." Fascinated by this shrub, "the Indian's rope," he
"cut a good-sized switch, which was singularly tough and
flexible, just like a cowhide," and went on about it in his

journal: "Frost said that the farmers of Vermont used it to tie up their fences with. Certainly there can be no wood equal to it as a withe. He says it is still strong when dry. I should think it would be worth the while for the farmers to cultivate for this purpose. How often in the woods and fields we want a string or rope and cannot find one. This is the plant which Nature has made for this purpose."

Of Charles Christopher Frost (1805–1881), botanist and shoemaker of Brattleboro, whom he knew, Thoreau tells this anecdote: "In the last heavy rain, two or three weeks since, there was a remarkable freshet on this brook [Whetstone], such as has not been known before, the bridge and road carried away, the bed of the stream laid bare, a new channel being made, the interval covered with sand and gravel, and trees (buttonwood, etc.) brought down; several acres thus buried. Frost escaped from his house on a raft."

Thoreau had read about leatherwood in Jacob Bigelow's *American Medical Botany*, but this was his only encounter with it. Though *Dirca* occurs in much of New England, it is absent from Concord. Had Thoreau time, when in Burlington, he might have found it in the Champlain valley, I was told by Robert Popp, head of Vermont's Heritage Program. I've yet to see it in Brattleboro, though ecologist Tom Wessels showed it to me on a ridgetop in Putney.

His last full day in Brattleboro, Thoreau crossed the Connecticut River and climbed Wantastiquet with Frances and Mary Brown, the eldest (age twenty-two) and the youngest (age fourteen) of the five Brown children. "High up the mountain the *Aster macrophyllus* as well as *corymbosus* [*divaricatus*].... Top of the mountain covered with wood. Saw Ascutney, between forty and fifty miles up the river, but not Monadnock on account of woods."

Both these asters, large-leaved and whitewood are still prevalent. Enduring fires and gypsy moths, the mountain's lush forest prevails. Mountain laurel brightens a mid-June day. The best views are west and south.

TEXT: *The Journal of Henry D. Thoreau,*
5–9 September 1856, IX:61–74.

Monadnock can be seen from several points in Brattleboro and environs. The Kiplings—Rudyard married Caroline Balestier of Brattleboro—lived in Brattleboro from 1892 to 1896. From their home, Naulakha, they viewed Monadnock "beyond the very furthest range," Kipling wrote, "where the pines turn to a faint blue haze against the one solitary peak—a real mountain and not a hill—showed like a gigantic thumbnail pointing heavenward."

Try taking a Thoreauvian saunter. From Brattleboro railroad station, walk over the Connecticut River on two bridges joined by an island. Turn left onto Mountain Road to the Wantastiquet Trail, which follows the former carriage road to the summit. A path beside the mountain along the river is also pleasant. Board the train the next day and follow Thoreau north to Bellows Falls (see Fall Mountain).

After receiving blooming bloodroot and mayflowers (trailing arbutus) from Mary Brown on 23 April 1858, Thoreau thanked her: "Please remember me to Father and Mother, whom I shall not fail to visit whenever I come to Brattleboro, also to the Chesterfield mountain, if you can communicate with it; I suppose it has not budged an inch" (*C,* 510).

Though Thoreau never met Wantastiquet again, bloodroot's white flowers still signal spring in this part of the world.

Fall

♦

After climbing Wantastiquet, Thoreau went north by train (23 miles) from Brattleboro to Bellows Falls. This lovely ride along the Connecticut River, with its glimpses of Fall Mountain (1115'), can still be done today.

From the Bellows Falls railroad station, passengers could go by stage up the Saxton's River valley 11.5 miles to the Tavern in Grafton, Vermont. When I visited in 1987, I was surprised to find Thoreau's name among the Tavern's illustrious guests printed over the registration desk and in its brochure, a rumor perpetuated in a popular guidebook. After a long search — the guest registers of this period had burned — I've not been able to place Thoreau in Grafton.

Instead, Thoreau walked from the depot in Bellows Falls, which still occupies the same site, past Island House on the corner of Central and Bridge streets, a grand hotel (open 1851–1887; no longer there) managed by C. R. White, whose guests went by carriage to the top of Fall Mountain. Thoreau paid the toll to cross the covered bridge, inspected the falls and potholes of the Connecticut River, and in New Hampshire "ascended the Fall Mountain with a heavy valise on my back, against the advice of the toll-man. But when I got up so soon and easily I was amused to remember his anxiety. It is seven hundred and fifty feet high, according to Gazetteer [John Hayward, *The New England Gazetteer* (1839)]. Saw great red oaks on this hill, particularly tall, straight, and bare of limbs, for a great distance, amid the woods. Here, as at Brattleboro, a fine view of the country immediately beneath you; but these views lack breadth, a distant horizon. There is a complete view of the falls from this height."

Since there is no guidebook, you can follow my directions. It is 1.5 miles from the depot to the base of the mountain. Walk or bicycle south on Island Street, turn left (east) onto Bridge Street over the river, whose flow is diverted for energy to the canal (1802), the first in America. Next, turn left (north) on Route 12 to North Walpole, right onto Main Street, then right onto Mountain View Road. Go uphill past the elementary school to the road's end. From here, take the dirt track—formerly a carriage road—for half a mile, then the path (right, south) to the lookout at Table Rock. Red and scrub oaks are plentiful here. Smell the lovely roseshell azalea (*Rhododendron roseum*) in early June.

"Descending the steep south end of this hill....My shoes were very smooth, and I got many falls descending, battering my valise." Having gone down the trailless south slope partly on my bottom, I recommend caution or retracing your steps.

Had Thoreau been here thirteen days later, on the afternoon of 23 September 1856, he would have met the distinguished geologist Edward Hitchcock, William A. Stearns, the president of Amherst College, along with twenty-nine members of Amherst's class of 1857, students from Middlebury and Dartmouth, invited guests, and the Bellows Falls band, who had climbed Fall Mountain to name it in honor of John Kilburn, Walpole's first white settler. Though objections were voiced at the ceremony, Mount Kilburn still identifies the southern peak of Fall Mountain.

Walpole

Propelled from Fall Mountain onto the road, now Route 12, Thoreau crossed Cold River, bathed in the Connecticut, then walked south for 2 miles, riding "the last mile into Walpole with a lumberer."

He knocked on the door of the Alcotts, who had moved to Walpole in July 1855. Bronson Alcott had invited him. The Alcott home was on the common where the library now stands, historian Guy Bemis told me when I visited in 1984; it had been moved to High Street, the second house (Walpole does not use numbers) on the north side. It was owned by Benjamin Willis, who married Abigail May Alcott's sister, Elizabeth.

"*Sept.* 11. P.M. — Walked over what Alcott calls Farm Hill, east of his house.

"*Erigeron annuus* [daisy fleabane], four feet high, by roadside; also *Ranunculus Pennsylvanicus* [*sic*], or bristly crowfoot, still in bloom. *Vide* press. A fine view of the Connecticut valley from the hilltop, and of Ascutney Mountain, but not of Monadnock. Descended a steep side of the hill by a cow-path, made with great judgment regularly zigzag, thus: well worn and deep."

"There never was a Farm Hill," historian Virginia Putnam writes; it was the general designation for the uplands around Walpole's Main Street where most of the mid-nineteenth-century farms were located. I was directed to Wentworth Hill, about half a mile south (Thoreau's course was east) of the library, on Wentworth Road, home of Henry and Frances Francis (Knapp House, c. 1812). Henry walked me up their back yard, a gradual incline through an open field of hawkweeds, daisies, buttercups, and red clover, to a wood-rail fence at the top from where I could see Fall Mountain and the Connecticut River but not Ascutney. He told me that Guy Bemis had brought Professor William Howarth here to show him "Mr. Thoreau's View." Later, while we drank tea on the porch, Frances read from *Under the Lilacs* by Louisa May Alcott, a first edition given to her by her father, who had received it from his grandmother in 1879, the setting for which, she explained, was this very ground. "Are the lilacs still here?" I asked. "Are they ever," she said, pointing over my shoulder. "All along the road." Henry then walked me across the lawn, away from the road into the woods, where a path led us to the head of a ravine with a spring-fed brook. "One of Louisa May Alcott's haunts," he said.

While here, you can walk the Academy Ravine to Louisa May Alcott Falls. Begin on School Street, to the left of the Walpole Highway Department.

TEXT: *The Journal of Henry D. Thoreau,*
10–12 September 1856, IX:74–80.

Monadnock

♦

To distinguish this from other Monadnocks, such as the one that rises along the Connecticut River in Vermont's Northeast Kingdom, which Thoreau never climbed, it is called the Grand Monadnock (3165'). Geologists use its name to identify similar isolated alps, such as Vermont's Ascutney (3150'), which Thoreau saw 44 miles northwest of Monadnock. This, Thoreau's most cherished mountain, the one he spent the most time with, is a National Natural Landmark.

"We proceeded to get our tea on the summit, in the very place where I had made my bed for a night some fifteen years before." On his third visit to Monadnock, actually fourteen years later (3 June 1858), Thoreau related only this about his first ascent in July 1844.

6–7 SEPTEMBER 1852

Temple Mountain

Eight years later, in 1852, Thoreau returned to Monadnock. On September 6 he went by train from Concord to

Mason Village (now Greenville), New Hampshire. "Walked from Mason village over the *Mt tops* to Peterboro.... High blackberries by the roadside abundant still —the long sweet mulberry shaped ones—mostly confined to the road—& very grateful to the walker." In 5 miles he stopped at a tavern in Temple, now the Birchwood Inn, whose brochure claims that Thoreau spent the night (he moved on to Peterborough before resting).

The innkeeper, Elias Colburn, informed Thoreau that the summit was just south of Peterborough Road (Route 101). "Went across lots from here toward this—when part way up—or on a lower part of the ridge—discovered it was not the highest & turned northward across the road to what is apparently the highest—first having looked south to Kidders *Mt* between New Ipswich & Temple & further west & quite near to Boundary Mt between Sharon & Temple— Already we had had experience of a *mt* side covered with bare rocks as if successive thunder spouts had burst over it—& bleached timber lying across the rocks—the woodbine red as blood about a tall stump—& the strong sweet bracing scent of ferns between the rocks—the raspberry bushes still retaining a few berries— They usually tell you how many *mt* houses you can see from a *mt*—but they are interesting to me in proportion to the *no* you cannot see."

Temple Mountain, due west of Temple village, culminates in Holt Peak (2084'). From my ramblings here, I concluded that Thoreau attained somewhere on Whitcomb Peak (1710'; called Fuller Mountain on old USGS maps), an outlier of spine of Temple Mountain, ¾ mile east of Holt Peak. What was summer pasture for Massachusetts cattle is now forest. Leaving Route 45 north of Temple, I pursued cross-country ski trails, woods road, path, and bushwhack to Whitcomb, which is covered with birch, beech, striped maple, hobblebush, stone walls, blueberries. The current owner, Edward Leighton, told me he had burned over Whitcomb to improve the blueberry yield. Kidders and Boundary mountains are now Kidder Mountain and Burton Peak, respectively.

Pack Monadnock

"We went down the west side of this first *mt* from whose summit we could not see W on account of another ridge — descended far & across the road & up the Southern most of what I have called the Peterboro hills — The raw edge of a forest of canoe birches on the side of this hill was remarkable — on account of the wonderful contrast of the white stems with the green leaves — the former glaringly white as if white washed & varnished or polished."

Instinct and various ski trails will guide you to Route 101, where you will see the ski center. Cross 101 into General James Miller State Park. On 17 September 1849, Hawthorne visited James Miller at his home a mile east of Temple center (look for the historic marker). They had met in Salem, where Miller served as collector of customs from 1825 to 1849. Hawthorne celebrated him in "The Custom-House — Introduction" of *The Scarlet Letter*. The next day, with his host's son Ephraim, Hawthorne rode "three or four miles, to the summit of an intervening ridge" from where Monadnock was visible.

The "Peterboro hills" comprise Pack Monadnock Mountain (2286'; South Pack) and North Pack Monadnock (2278'), which are joined by the Wapack Trail. Atop

South Pack, Thoreau saw "Lyndeboro *mt* north of these two...& further Crotched *mt*—and in the N.E. Uncan-nunuc." He had climbed the latter on 5 September 1848. From the summit tower I could see Thoreau's route from Greenville to here.

"Descended where as usual the forest had been burned formerly—tall bleached masts still standing— making a very wild & agreeably [*sic*] scenery.—keeping on a westward spur or side that we might see N & S— Saw the pond [Topside] on the 'embenchement' be-tween the two *mts.* Some sheep ran from us in great fear. Others put their heads down & together & stood *per-fectly still*—resembling rocks—so that I did not notice them at first— Did they not do it for concealment?"

You can follow Thoreau north on the Wapack Trail. Descend for .5 mile into the col, then bushwhack down and west to Topside Pond, which I reached no less elated than Balboa at the Pacific. After the owners let me swim, I took their beautiful long drive to Old Mountain Road, turned south to the first right (west), now Carley Road, to Old Street Road, then left and then right onto Cheney Street to Pine Street, which led me north into the village of Peterborough at 7 P.M. Now the 1.6-mile Raymond Trail connects Pack Monadnock with Old (now East) Mountain Road, facilitating this descent.

Thoreau might have stayed at one of two public houses in Peterborough: the one on Main Street, where the Granite Bank is now, or the one at the corner of Union and Elm, the Upper Hotel, which has been con-verted to apartments.

"A man in Peterboro told me that his father told him that Monadnock used to be covered with forest—that

fires ran through it & killed the turf—then the trees were
blown down & their roots turned up & formed a dense
& impenetrable thicket in which the wolves abounded—
They came down at night killed sheep &c & returned to
their dens whither they could not be pursued, before
morning—till finally they set fire to this thicket & it
made the greatest fire they had ever had in the county—
& drove out all the wolves which have not troubled them
since."

Tuesday, 7 September, was a colossal day for Thoreau. He walked from Peterborough, via Joe Eveleth's
house, to the summit of Monadnock by one o'clock, a
distance of 6 or 7 miles, he estimated, pausing to
botanize:

"In one little hollow between the rocks grew—blue
berries—chokeberries—bunch berries——*red* cherries
—wild currants (ribes prostratum with the berry the
odor of skunk cabbage—but a not quite disagreeable
wild flavor) a few raspberries still—holly-berries—*mt*
cranberries (*Vaccinium vitis idaea*) all close together.
The little soil on the summit between the rocks was covered with the Potentilla tridentata now out of bloom—
the prevailing plant at the extreme summit. Mt ash
berries also."

He coasted down the south side of Monadnock and
into Troy, dashing to catch the 3 P.M. train to Concord.
He was home at 5:15, "4 hours from the time we were
picking blueberries on the *mt*—with the plants of the *mt*
fresh in my hat."

When I retraced Thoreau, I left Peterborough center
at 8:47 A.M., walked 8 miles, and reached the Eveleth
House at the end of Burpee Road in Dublin at 1:20 P.M.

The owners, Bruce and Mary Elizabeth McClellan, said that there was no path from here on and suggested I go by compass, which I did not have. Laughingly, they estimated my time of arrival in Troy at 3 A.M. Up through the woods I went, thinking they might be right, until I met the Pumpelly Trail, which I followed to the top of Monadnock, arriving at 4:45. Immediately I descended the White Arrow Trail to the Old Toll Road to Route 124, then west to Monadnock Street and south to the extant Troy station, a block southwest of the common, arriving at 7:15. My results: 17 miles in almost 10½ hours. Thoreau's walk may have been slightly shorter, since he moved in a straight line through relatively open terrain. If he had left at dawn, 4:30 A.M., which is in character, though his departure time and place are unknown, his travel time would have been the same. These long two days gave me an appreciation for Thoreau's physical energy.

TEXT: *Henry D. Thoreau: Journal 5, 1852–1853,* 336–40.

2–4 JUNE 1858

It was six years before Thoreau returned to Monadnock. This time in late spring; this time to stay for three days and write extensively in his journal; this time with his Worcester friend H. G. O. Blake. They arrived by train in Troy, New Hampshire, "at 11.5 and shouldered our knapsacks, steering northeast to the mountain, some four miles off, — its top....

"Almost without interruption we had the mountain in sight before us, — its sublime gray mass — that antique, brownish-gray, Ararat color. Probably these crests of the earth are for the most part of one color in all lands, that gray color of antiquity, which nature loves."

Ararat (16,945'), in eastern Turkey, the assumed landing place of Noah's Ark, is also the name of a Cape Cod sand dune Thoreau had ascended a year before, on 21 June 1857, as well as on 13 October 1849. Later he discovered that the dun color resulted from the presence of brown lichens.

"We left the road at a schoolhouse, and, crossing a meadow, began to ascend gently through very rocky pastures." The schoolhouse, at the southeast corner of Monadnock Street and Route 124, is gone.

"Entering the wood we soon passed Fassett's shanty, —he so busily at work inside that he did not see us,— and we took our dinner by the rocky brook-side in the woods just above."

Fassett's stood just beyond the former site of the Halfway House, 1.2 miles from Route 124 on the Old Toll Road, which was cut by Joseph Fassett, whom Thoreau saw. A stone marker with "Mountain House JF 1857" remains. Thereabouts Thoreau observed in bloom skunk currant, hobblebush, large-flowered bellwort, red elderberry "not open," smooth shadbush, and "the handsomest flower of the mountain," painted trillium.

In midafternoon they found a campsite—"a sunken yard in a rocky plateau on the southeast side of the mountain, perhaps half a mile from the summit"—and prepared a shelter, which Thoreau described. Then they proceeded to the top "to see the sun set." There he discovered slate-colored juncos and their nest with three eggs. "We saw many of these birds flitting about the summit, perched on the rocks and the dwarf spruce, and disappearing behind the rocks. It is the prevailing bird now . . . on the summit." And "the prevailing plants of a high order" he noted — mountain cranberry, mountain sandwort, three-toothed cinquefoil — can still be seen.

"As it was quite hazy, we could not see the shadow of the mountain well, and so returned just before the sun set to our camp. We lost the path coming down, for nothing is easier than to lose your way here, where so little trail is left upon the rocks, and the different rocks and ravines are so much alike. Perhaps no other equal area is so bewildering in this respect as a rocky mountain-summit, though it has so conspicuous a central point."

Thoreau and Blake kindled a fire from dead spruce to boil spring water for "taking our tea in the twilight." Nighthawks serenaded them in their spruce bed, "their dry and unmusical, yet supramundane and spirit-like, voices and sounds gave fit expression to this rocky mountain solitude. It struck the very key-note of the stern, gray, barren solitude. It was a thrumming of the mountain's rocky chords."

Up at 3:30 A.M., "in order to see the sun rise from the top and get our breakfast there." After which he inventoried plants, then turned to geology. "We observed that the rocks were remarkably smoothed, almost polished and rounded, and also scratched. The scratches run from about north-northwest to south-southeast. . . . There were occasionally conspicuous masses and also veins of white quartz, and very common were bright-purple or wine-colored garnets imbedded in the rock, looking like berries in a pudding. . . .

"The little bogs or mosses, sometimes only a rod in

diameter, are a singular feature. Ordinarily the cladonia and other lichens are crackling under your feet, when suddenly you step into a miniature bog filling the space between two rocks and you are at a loss to tell where the moisture comes from....

"The southeast part of the mountain-top is an extended broad rocky *almost* plateau, consisting of large flat rocks with small bogs and rain-water pools and easy ascents to different levels. The black spruce tree which is scattered here and there over it, the prevailing tree or shrub of the mountain-top, evidently has many difficulties to contend with." Actually, red spruce, *Picea rubens* (not distinguished as a separate species in Thoreau's day), covered Monadnock until the fires of 1800 and 1820.

"We boiled some rice for our dinner, close by the edge of a rain-water pool and bog, on the plateau southeast from the summit. Though there was so little vegetation, our fire spread rapidly through the dry cladonia lichens on the rocks, and, the wind being pretty high, threatened to give us trouble, but we put it out with a spruce bough dipped in the pool....

"After dinner we kept on northeast over a high ridge east of the summit, whence was a good view of that part of Dublin and Jaffrey immediately under the mountain. There is a fine, large lake [Thorndike Pond] extending north and south, apparently in Dublin, which it would be worth the while to sail on." On this side of the mountain Thoreau saw bogs, still living, one of which, at the headwaters of Mountain Brook, is named for him. After surveying the mountain, Thoreau reflected: "It often reminded me of my walks on the beach, and suggested how much both depend for their sublimity on solitude and dreariness. In both cases we feel the presence of some vast, titanic power. The rocks and valleys and bogs and rain-pools of the mountain are so wild and unfamiliar still that you do not recognize the one you left fifteen minutes before. This rocky region, forming what you may call the top of the mountain, must be more than two miles long by one wide in the middle, and you would need to ramble about it many times before it would begin to be familiar."

On Friday, June 4, they started down. "It is remarkable how, as you are leaving a mountain and looking back at it from time to time, it gradually gathers up its slopes and spurs to itself into a regular whole, and makes a new and total impression.... Even at this short distance the mountain has lost most of its rough and jagged outline, considerable ravines are smoothed over, and large boulders which you must go a long way round make no impression on the eye, being swallowed up in the air."

They walked south through west Rindge to State Line Station, a 10-mile beeline. From there, following the railroad south, Thoreau "saw some of the handsomest white pines here that I ever saw.... Their regular beauty made such an impression that I was forced to turn aside and contemplate them." In 3 miles they came to Winchendon, Massachusetts, from where they entrained for Concord.

TEXT: *The Journal of Henry D. Thoreau,* x:452–80.

4–9 AUGUST 1860

Thoreau's last communion with Monadnock was his longest. Repeating his approach of two years before, he and Ellery Channing went by train to Troy, New Hampshire, and from there afoot, 3 miles along Monadnock Street to Route 124, "exhilarated by the peculiar raspberry scent by the roadside this wet day—and of the dicksonia fern." They passed what is now called The Inn at East Hill Farm.

"We crossed the immense rocky and springy pastures, containing at first raspberries, but much more hardhack in flower, reddening them afar, where cattle and horses collected about us, sometimes came running to us, as we thought for society, but probably not....

"We were wet up to our knees before reaching the woods or steep ascent where we entered the cloud. It was quite dark and wet in the woods, from which we emerged into the lighter cloud about 3 P.M., and proceeded to construct our camp, in the cloud occasionally amounting to rain, where I camped some two years ago." Thoreau built a spruce shelter with a fire at its door, which dried them completely in an hour or two.

The next day was clear. "The rocks of the main summit were olive-brown, and C. called it the Mount of Olives.

"I had gone out before sunrise to gather blueberries, —fresh, dewy (because wet with yesterday's rain), almost crispy blueberries, just in prime, much cooler and more grateful at this hour,—and was surprised to hear the voice of people rushing up the mountain for berries in the wet, even at this hour. These alternated with bright light-scarlet bunchberries not quite in prime.

"The sides and angles of the cliffs, and their rounded brows ... were clothed with these now lively olive-brown lichens (umbilicaria), alike in sun and shade, becoming afterward and generally dark olive-brown when dry."

He walked "to the lower southern spur of the mountain...and the whole mountain-top for two miles was covered, on countless little shelves and in hollows between the rocks, with low blueberries of two or more species or varieties, just in their prime...."

"These blueberries grew and bore abundantly almost wherever anything else grew on the rocky part of the mountain... quite up to the summit, and at least thirty or forty people came up from the surrounding country this Sunday to gather them. When we behold this summit at this season of the year, far away and blue in the horizon, we may think of the blueberries as blending their color with the general blueness of the mountain."

Mountain cranberries also captivated him. "We gathered a pint in order to make a sauce of them.... We stewed these berries for our breakfast the next morning, and thought them the best berry on the mountain, though, not being quite ripe, the berry was a little bitterish—but not the juice of it. It is such an acid as the camper-out craves."

The next day, August 6, they "moved about a quarter of a mile along the edge of the plateau eastward and built a new camp there" from where they "could look all over the south and southeast world without raising our heads."

He took an afternoon saunter to northeast swamps, now bogs. "These two swamps are about the wildest part of the mountain and most interesting to me. The smaller occurs on the northeast side of the main mountain, *i.e.* at the northeast end of the plateau. It is a little roundish meadow of a few rods over, with cotton-grass in it, the shallow bottom of a basin of rock, and out the east side there trickles a very slight stream, just moistening the rock at present and collecting enough in one cavity to afford you a drink....

"The larger swamp is considerably lower and more northerly, separating the northeast spur from the main mountain, probably not far from the line of Dublin. It extends northwest and southeast some thirty or forty rods, and probably leaked out now under the rocks at the northwest end.... The prevailing grass or sedge in it, growing in tufts in the green moss and sphagnum between the fallen dead spruce timber, was the *Eriophorum vaginatum* [hare's-tail, *E. spissum*] (long done) and *E. gracile* [slender cotton-grass]. Also the *Epilobium palustre* [marsh willow-herb], apparently in prime in it, and common wool-grass (*Scirpus eriophorum*). Around its edge grew the *Chelone glabra* [turtlehead] (not yet out), and meadow-sweet in bloom, black choke-berry just ripening, red elder (its fruit in prime), mountain-ash, *Carex trisperma* [three-seeded sedge] and *Deweyana* [Dewey's sedge] (small and slender), and the fetid currant in fruit (in a torrent of rocks at the east end), etc., etc.

"I noticed a third..." When botanist Robert Buchsbaum and I explored these three bogs in 1987, Bob discovered sundews, *Drosera rotundifolia,* in Thoreau's Bog, the largest of the three and the first we encountered going west along Pumpelly Trail toward the summit. We named the second bog Cotton-grass, also north of the trail, though closer, and west of Sarcophagus Rock, and the third bog, this one south of Pumpelly Trail, we called

Leatherleaf (Thoreau's andromeda, familiar to him from Concord), after the distinctive plants of these unnamed bogs.

Back at camp after five, Thoreau's gaze went outward. "Evening and morning were the most interesting seasons, especially the evening. Each day, about an hour before sunset, I got sight, as it were accidentally, of an elysium beneath me. The smoky haze of the day, suggesting a furnace-like heat, a trivial dustiness, gave place to a clear transparent enamel, through which houses, woods, farms, and lakes were seen as in [a] picture indescribably fair and expressly made to be looked at. At any hour of the day, to be sure, the surrounding country looks flatter than it is. Even the great steep, furrowed, and rocky pastures, red with hardhack and raspberries, which creep so high up the mountain amid the woods, in which you think already that you are half-way up, perchance, seen from the top or brow of the mountain are not for a long time distinguished for elevation above the surrounding country, but they look smooth and tolerably level, and the cattle in them are not noticed or distinguished from rocks unless you search very particularly. . . .

"But, above all, from half an hour to two hours before sunset many western mountain-ranges are revealed, as the sun declines, one behind another. . . . I could count in the direction of Saddleback Mountain [Greylock, 56 miles west] eight distinct ranges, revealed by the darker lines of the ridges rising above this cloud-like haze. And I might have added the ridge of Monadnock itself within a quarter of a mile of me. . . .

"I never saw a mountain that looked so high and so melted away at last cloud-like into the sky, as Saddleback this eve, when your eye had clomb to it by these eight successive terraces. You had to begin at this end and ascend step by step to recognize it for a mountain at all. If you had first rested your eye on *it,* you would have seen it for a cloud, it was so incredibly high in the sky....

"*Aug.* 7.... I rose always by four or half past four to observe [sunrise]. The sun rose about five....

"There was every morning more or less solid white fog to be seen on the earth, though none on the mountain. I was struck by the localness of these fogs. For five mornings they occupied the same place and were about the same in extent. It was obvious that certain portions of New Hampshire and Massachusetts were at this season commonly invested with fog in the morning, while others, or the larger part, were free from it. The fog lay on the lower parts only. From our point of view the largest lake of fog lay in Rindge and southward; and southeast of Fitzwilliam, *i.e.* about Winchendon, very large there. In short, the fog lay in great spidery lakes and streams answering to the lakes, streams, and meadows beneath, especially over the sources of Miller's River and the region of primitive wood thereabouts....

"*Aug.* 8. *Wednesday.* 8.30 A.M. Walk round the west side of the summit. Bathe in the rocky pool there, collect mountain cranberries on the northwest side, return over the summit, and take the bearings of the different spurs, etc." (see map).

"*Aug.* 9. At 6 A.M., leave camp for Troy, where we arrive, after long pauses, by 9 A.M., and take the cars at 10.5."

So concluded his journey, but not his journal. Thoreau wrote on, listing summit plants, discussing spruce, shrubs, birds, quadrupeds, insects, frogs, and humans:

"There were a great many visitors to the summit, both by the south and north, *i.e.* the Jaffrey and Dublin paths, but they did not turn off from the beaten track. One noon, when I was on the top, I counted forty men, women, and children around me, and more were constantly arriving while others were going. Certainly more than one hundred ascended in a day. When you got within thirty rods you saw them seated in a row along the gray parapets, like the inhabitants of a castle on a gala-day; and when you behold Monadnock's blue summit fifty miles off in the horizon, you may imagine it covered with men, women, and children in dresses of all colors,

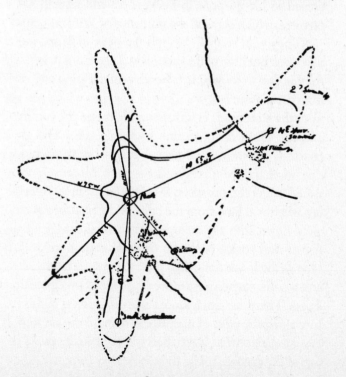

SCALE OF 80 RODS TO AN INCH.

like an observatory on a muster-field. They appeared to be chiefly mechanics and farmers' boys and girls from the neighboring towns. The young men sat in rows with their legs dangling over the precipice, squinting through spy-glasses and shouting and hallooing to each new party that issued from the woods below. Some were playing cards; others were trying to see their house or their neighbor's. Children were running about and playing as usual. Indeed, this peak in pleasant weather is the most trivial place in New England. There are probably more arrivals daily than at any of the White Mountain houses."

After relating Monadnock's geography, Thoreau returned to his purpose in being there and framed some pictures, which showed the influence of William Gilpin.

"They who simply climb to the peak of Monadnock have seen but little of the mountain. I came not to look *off from* it, but to look *at* it. The view of the pinnacle itself from the plateau below surpasses any view which you get from the summit. It is indispensable to see the top itself and the sierra of its outline from one side. The great charm is not to look off from a height but to walk over this novel and wonderful rocky surface. Moreover, if you would enjoy the prospect, it is, methinks, most interesting when you look from the edge of the plateau immediately down into the valleys, or where the edge of the lichen-clad rocks, only two or three rods from you, is seen as the lower frame of a picture of green fields, lakes, and woods, suggesting a more stupendous precipice than exists. There are much more surprising effects of this nature along the edge of the plateau than on the summit. It is remarkable what haste the visitors make to get to the top of the mountain and then look away from it."

Thoreau then concentrated on geology. "But what a study for rocks does this mountain-top afford!... You see huge buttresses or walls put up by Titans, with true joints, only recently loosened by an earthquake as if ready to topple down.... You see large boulders, left just on the edge of the steep descent of the plateau, commonly resting on a few small stones, as if the Titans were in the very act of transporting them when they were interrupted; some left standing on their ends, and almost the only convenient rocks in whose shade you can sit sometimes....

"The rocks were not only coarsely grooved but finely scratched from northwest to southeast."

Thoreau's Titans were the glaciers that covered Monadnock. Even the famed Edward Hitchcock, in his *Final Report on the Geology of Massachusetts,* Thoreau's authority, attributed this striking landscape to "diluvial" effects.

From geology, Thoreau went on to describe clouds: "Every evening, excepting, perhaps, the Sunday evening after the rain of the day before, we saw not long after sundown a slight scud or mist begin to strike the summit above us, though it was perfectly fair weather generally and there were no clouds over the lower country.

"First, perhaps, looking up, we would see a small scud not more than a rod in diameter drifting just over the apex of the mountain. In a few minutes more a somewhat larger one would suddenly make its appearance, and perhaps strike the topmost rocks and invest them for a moment, but as rapidly drift off northeast and disappear. Looking into the southwest sky, which was clear, we would see all at once a small cloud or scud a rod in diameter beginning to form half a mile from the summit, and as it came on it rapidly grew in a mysterious manner, till it was fifty rods or more in diameter, and draped and concealed for a few moments all the summit above us, and then passed off and disappeared northeastward just as it had come on. So that it appeared as if the clouds had been attracted by the summit. They also seemed to rise a little as they approached it, and endeavor to go over without striking. I gave this account of it to myself. They were not attracted to the summit, but simply generated there and not elsewhere. There would be a warm southwest wind blowing which was full of moisture, alike over the mountain and all the rest of the country. The summit of the mountain being cool, this warm air began to feel its influence at half a mile distance, and its moisture was rapidly condensed into a small cloud, which expanded as it advanced, and evaporated again as it left the summit. This would go on, apparently, as the coolness of the mountain increased, and generally the cloud or mist reached down as low as our camp from time to time, in the night.

"One evening, as I was watching these small clouds forming and dissolving about the summit of our mountain, the sun having just set, I cast my eyes toward the dim bluish outline of the Green Mountains in the clear red evening sky, and, to my delight, I detected exactly over the summit of Saddleback Mountain, some sixty miles distant, its own little cloud, shaped like a parasol and answering to that which capped our mountain, though in this case it did not rest on the mountain, but

was considerably above it, and all the rest of the west horizon for forty miles was cloudless.

"I was convinced that it was the local cloud of that mountain because it was directly over the summit, was of small size and of umbrella form answering to the summit, and there was no other cloud to be seen in that horizon. It was a beautiful and serene object, a sort of fortunate isle, — like any other cloud in the sunset sky."

He closed with lists — the contents of his knapsack and his provisions for six days: 18 hard-boiled eggs, 2½ pounds of sugar, salt, ¼ pound of tea, two pounds of hard bread, half a loaf of homemade bread, and a piece of cake. He noted for next time to omit eggs, reduce the sugar by half a pound and the tea by a third, bring the same amount of bread, and "more home-made and more *solid* sweet cake."

TEXT: *The Journal of Henry D. Thoreau*, XIV:8–52.

"We & the *mt* had a sound season," Thoreau wrote H. G. O. Blake on 4 November 1860, adding this story about his and Channing's excursion: "An old Concord farmer tells me that he ascended Monadnock once, & danced on the top. How did that happen? Why, he being up there, a party of young men & women came up bringing boards & a fiddler, and having laid down the boards they made a level floor, on which they danced to the music of the fiddle. I suppose the tune was 'Excelsior'" (*C*, 597). Every August since 1985, Dianne Eno has choreographed dance performances for the summit, executing miracles on seemingly impossible ground.

In the twentieth century, Monadnock continued to attract artists and writers. Willa Cather summered at the Shattuck Inn in Jaffrey from 1917 to 1937. Each summer from 1911 to 1934 Edwin Arlington Robinson wrote poems at the MacDowell Colony in Peterborough, which has nurtured creativity since 1907. English watercolorist Tony Foster, whom I guided in New Hampshire and Maine, ascended Monadnock and Skatutakee Mountain in the mid-1980s to paint for his "Thoreau's Country" exhibit at the Yale Center for British Art.

Keene

Thoreau's Canada-bound train stopped in Keene on 25 September 1850. The depot was adjacent to the house where his mother, Cynthia Dunbar, was born, a month before the death of her father, who had built the house in 1785. Her residence until 1798, though highly modified, still stands at 81 Main Street. Though Keene is visible from Monadnock, Thoreau did not visit it on his last three Monadnock excursions. It is possible, however, that he walked through Keene on his way west in July 1844, though more likely he would have gained access to the Ashuelot River valley south of Keene, which I believe he followed to the Connecticut River.

*I think that the top of Mt. Washington should not be
private property; it should be left unappropriated for
modesty and reverence's sake, or if only to suggest
that earth has higher uses than we put her to.*

3 January 1861

Washington

♦

Thoreau made two excursions to Mount Washington
(6288'). On the first, from 31 August to 13 September
1839, he went with his brother John by boat on the Concord and Merrimack rivers to Hooksett, from where they
walked 12 miles to Concord and spent the night (place
unknown). The next day they went by stage to Plymouth,
thence afoot 13 miles to Tilton's Inn in West Thornton,
having their first view of the White Mountains from San-
bornton. The following day they walked 20 miles
through Franconia Notch, taking in the wonders of the
Flume, the Basin, and the Old Man of the Mountain, and
staying in Franconia (place unknown, though I suggest
Lovett's, 2 miles south of the hamlet on Route 18). From
Franconia they went on 20.5 miles to Thomas J. Crawford's for two nights, September 8 and 9. Thomas Crawford's place, called the Notch House, at Gateway of the
Notch (north end) opposite Saco Lake, was built in 1828
and burned in 1854. It is shown in Thomas Cole's famous oil *The Notch of the White Mountains (Crawford
Notch),* which Cole sketched two months before
Thoreau arrived. On the tenth Thoreau and his brother

ascended Washington following the Crawford Path 8.2 miles to the summit and back in time to ride to Conway —probably in a Crawford wagon—for the night. This path, laid out in 1819 by Abel Crawford and his son Ethan Allen, still serves ramblers. Celebrating the 150th anniversary of Thoreau's ascent, I spent ten and a half hours replicating it. Granted, Thoreau was twenty-seven years younger than I, but this 16.4-mile round-trip would challenge a climber of any age. In five days the Thoreaus had walked almost 82 miles.

TEXT: *Journal*, 1:134–37; *A Week on the Concord and Merrimack Rivers.*

2–19 JULY 1858

He returned nineteen years later, this time with his Maine companion of the previous summer, Edward Hoar, a Concord attorney. Traveling northward in a private carriage their first day, Thoreau found the Merrimack River inspiring: "This vista was incredible there. Suddenly I see a broad reach of blue beneath, with its curves and headlands, liberating me from the more terrene earth. What a difference it makes whether I spend my four hours' nooning between the hills by yonder roadside, or on the brink of this fair river, within a quarter of a mile of that! Here the earth is fluid to my thought, the sky is reflected from beneath, and around yonder cape is the highway to other continents. This current allies me to all the world. Be careful to sit in an elevating and inspiring place. There my thoughts were confined and trivial, and I hid myself from the gaze of travellers. Here they are expanded and elevated, and I am charmed by the beautiful river-reach."

They spent the night in a tavern in Merrimack. "The wood thrush sings almost wherever I go, eternally reconsecrating the world, morning and evening, for us."

Sunday, 4 July: "Leaving Loudon Ridge on the right we continued on by the Hollow Road—a long way through the forest without houses—through a part of Canterbury into Gilmanton Factory village."

Walking this way on 4 July 1985, I stopped at the Shaker Village in Canterbury for a tour of their buildings and camped in their parking lot. Thoreau's track, Hollow Road (unidentified by H. F. Walling, *Map of Merrimack County*, 1858), appears to be Route 106 now, which parallels Shaker Road, 1.25 miles east of Shaker Village. It's curious that Thoreau came so close to the Shakers yet did not visit or even refer to them. Emerson had been here twice, on 2 January 1828 and 7–8 August 1829. Factory Village was part of Gilmanton until 1859, when it became Upper Gilmanton. Since 1869 it has been the town of Belmont.

"We continued along through Gilmanton to Meredith Bridge, passing the Suncook Mountain on our right, a long, barren rocky range overlooking Lake Winnepiseogee. Turn down a lane five or six miles beyond the bridge and spend the midday near a bay of the lake." Gunstock Mountain, one of the peaks of the Belknap Range, was locally called Suncook. Thoreau must not have realized that what he was seeing he had called Gunstock when he had viewed it from Uncanoonuc in 1848. Meredith Bridge was incorporated as the town of Laconia in 1855.

Red Hill

On 5 July they walked up Red Hill (2029'). "Dr. [Charles T.] Jackson says that Red Hill is so called from the uva-ursi [bearberry] on it turning red in the fall. On the top we boil a dipper of tea for our dinner and spend some hours, having carried up water the last half-mile.

"Enjoyed the famous view of Winnepiseogee and its islands southeasterly and Squam Lake on the west, but I was as much attracted at this hour by the wild mountain view on the northward. Chocorua and the Sandwich Mountains a dozen miles off seemed the boundary of cultivation on that side, as indeed they are." He essayed montane New Hampshire, possibly as far south as Monadnock, "dim and distant blue." This is a pleasant 1.7-mile trail.

When Emerson preceded the Thoreau brothers to the White Mountains, he ascended Red Hill on 26 August 1839 and met a Mrs. Cook, who had lived there for fifty-one years (*JMN*, 7:233–34).

"Descended, and rode along the west and northwest side of Ossipee Mountain.... We were all the afternoon riding along under Ossipee Mountain, which would not be left behind, unexpectedly large still, louring over your path. Crossed Bearcamp River, a shallow but unexpectedly sluggish stream, which empties into Ossipee Lake. Have new and memorable views of Chocorua, as we get round it eastward. Stop at Tamworth village for the night."

While courting Elinor White, Robert Frost spent the summer of 1895 on Mount Shaw (2990'), the apex of the

Ossipee Mountains. Castle in the Clouds, open to the public from mid-May to mid-October, now occupies the former site of Frost's residence.

Departing, 6 July, at 5:30 A.M., Thoreau and Hoar stopped for breakfast beside Silver Lake. North of here and west of Route 113 is Madison Boulder. "The scenery in Conway and onward to North Conway is surprisingly grand. You are steadily advancing into an amphitheatre of mountains."

The painter Benjamin Champney agreed with Thoreau when he stayed at the Kearsarge Tavern in North Conway. He painted Mount Washington from Sunset Hill, where, three years later in 1853, he bought a house, added a studio, and lived with his wife and family. This became a haven for artists. He was in Conway in July, and on the sixth when Thoreau passed through, he was sketching Mount Chocorua.

Champney, born the same year as Thoreau, was a charter member of the Boston Art Club and exhibited at the Boston Athenaeum. Thoreau knew three of his illustrations from Benjamin G. Willey's *Incidents in White Mountain History* (1856), which he had with him; the frontispiece is Champney's "White Mountains from North Conway," showing his farmhouse in the foreground. You can still see his home and studio on the corner of Main Street and Locust Lane.

"At Bartlett Corner we turned up the Ellis River and took our nooning on the bank of the river, by the bridge just this side of Jackson Centre, in a rock maple grove."

The next day, with William Wentworth, a local farmer, they "rode onward to the Glen House, eight miles further, sending back our horse and wagon to his house.

This road passes through what is called the Pinkham
Notch, in Pinkham's Grant, the land, a large tract, having
been given away to Pinkham for making the road a good
while since. Wentworth has lived here thirty years and is
a native." At 11:30 A.M., from 1600', they started up Mount
Washington via the Glen Bridle Path, which became the
carriage road and now is the 8-mile auto road. Thoreau
observed the changes of trees as they went higher, to "the
shanty where we spent the night with the colliers"—the
approximate former location of Halfway House (3840').

"I went on nearly a mile and a half further, and found
many new alpine plants and returned to this shanty. A
merry collier and his assistant, who had been making
coal for the summit and were preparing to leave the next
morning, made us welcome to this shanty and enter-
tained us with their talk. We here boiled some of our
beef-tongues, a very strong wind pouring in gusts down
the funnel and scattering the fire about through the
cracked stove. This man, named Page, had imported
goats on to the mountain, and milked them to supply us
with milk for our coffee. The road here ran north and
south to get round the ledge. The wind, blowing down
the funnel, set fire to a pile of dirty bed-quilts when I was
out, and came near burning up the building."

Early the next morning, 8 July, Thoreau walked to the
summit, sweating in a thick coat while taking in the
plants of the land above the trees: mountain sandwort,
Arenaria (his *Alsine*) *groenlandica;* Labrador tea, *Ledum
groenlandicum;* alpine bilberry, *Vaccinium uliginosum;*
bearberry willow, *Salix uva-ursi;* diapensia, *Diapensia
lapponica*—"Quite out of bloom; only one flower seen.
It grows in close, *firm,* and dense rounded tufts, just like

a moss but harder, between the rocks, the flowers considerably elevated above its surface." He also saw black crowberry, *Empetrum nigrum;* mountain cranberry, *Vaccinium vitis-idaea;* goldthread, *Coptis groenlandica* — with "dwarf shrubby canoe birches and almost impassable thickets of dwarf fir and spruce. The latter when dead exhibited the appearance of deer's horns, their hard, gnarled, slow-grown branches being twisted in every direction." He also noted trailing arbutus, *Epigaea repens;* stiff clubmoss, *Lycopodium annotinum;* fir clubmoss, *Lycopodium selago;* three-toothed cinquefoil, *Potentilla tridentata;* mountain avens, *Geum peckii* (his *radiatum* var. *peckii*); "and a little *Silene acaulis* (moss campion), still in bloom, a pretty little purplish flower growing like a moss in dense, hard tufts. . . .

"I got up about half an hour before my party and enjoyed a good view, though it was hazy, but by the time the rest arrived a cloud invested us all, a cool driving mist, which wet you considerably, as you squatted behind a rock."

When Thoreau stood here nineteen years earlier, the summit was wild. Now two stone dwellings confronted him. The Summit and Tip Top House offered meals and lodging and were managed as one by Joseph S. Hall and John H. Spaulding, both of whom Thoreau met. A restored Tip Top now serves as a museum.

Emerson went to the summit of Washington twice that I can document: on 16 July 1832 on foot over Crawford Path 2 from Ethan Allen Crawford's — preceding by two months Nathaniel Hawthorne by the same way — and on 5 September 1870 via carriage up and railway down.

"About 8:15 A.M., being still in a dense fog, we started

direct for Tuckerman's Ravine, I having taken the bearing of it before the fog, but Spaulding also went some ten rods with us and pointed toward the head of the ravine. ... I looked at my compass every four or five rods and then walked toward some rock in our course, but frequently after taking three or four steps, though the fog was no more dense, I would lose the rock I steered for. The fog was very bewildering....

"Descending straight by compass through the cloud, toward the head of Tuckerman's Ravine, we found it an easy descent over, for the most part, bare rocks, not very large, with at length moist springy places, green with sedge, etc., between little sloping shelves of green meadow, where the hellebore grew, within half a mile of top, and the *Oldenlandia caerulea* [bluets, *Houstonia caerulea*] was abundantly out (!) and very large and fresh, surpassing ours in the spring....

"We crossed a narrow portion of the snow, but found it unexpectedly hard and dangerous to traverse. I tore up my nails in my efforts to save myself from sliding down its steep surface. The snow-field now formed an irregular crescent on the steep slope at the head of the ravine, some sixty rods wide horizontally, or from north to south, and twenty-five rods wide from upper to lower side. It may have been half a dozen feet thick in some places, but it diminished sensibly in the rain while we were there. Is said to be all gone commonly by the end of August. The surface was hard, difficult to work your heels into, and a perfectly regular steep slope, steeper than an ordinary roof from top to bottom. A considerable stream, a source of the Saco, was flowing out from beneath it, where it had worn a low arch a rod or more wide. Here were the phenomena of winter and earliest spring, contrasted with summer. On the edge of and beneath the overarching snow, many plants were just pushing up as in our spring."

At the edge of the ravine they came into the sun again, which made visible at the bottom of the ravine Hermit Lake, toward which they steered. "But following down the edge of the stream, the source of Ellis River, which was quite a brook within a stone's throw of its head, we soon found it very bad walking in the scrubby fir and spruce, and therefore, when we had gone about two thirds the way to the lake, decided to camp in the midst of the dwarf firs, clearing away a space with our hatchets. Having cleared a space with some difficulty where the trees were seven or eight feet high, Wentworth kindled a fire on the lee side, without—against my advice—removing the moss, which was especially dry on the rocks and directly ignited and set fire to the fir leaves, spreading off with great violence and crackling over the mountain, and making us jump for our baggage; but fortunately it did not burn a foot toward us, for we could not have run in that thicket. It spread particularly fast in the procumbent creeping spruce, scarcely a foot deep, and made a few acres of deer's horns, thus leaving our mark on the mountainside....

"Finally we kept on, leaving the fire raging, down to the first little lake, walking in the stream, jumping from rock to rock with it. It may have fallen a thousand feet within a mile below the snow, and we camped on a slight rising ground between that first little lake and the stream, in a dense fir and spruce wood thirty feet high, though it was but the limit of trees there. On our way we found the

Arnica mollis [hairy arnica] (recently begun to bloom), a very fragrant yellow-rayed flower, by the side of the brook (also half-way up the ravine). The *Alnus viridis* [American green alder, *A. crispa*] was a prevailing shrub all along this stream, seven or eight feet high near our camp near the snow. It was dwarfish and still in flower, but in fruit only below; had a glossy, roundish, wrinkled, green, sticky leaf. Also a little *Ranunculus abortivus* [small-flowered buttercup] by the brook, in bloom.

"Close by our camp, the *Heracleum lanatum,* or cow-parsnip, masterwort, grew quite rankly, its great leaves eighteen inches wide and umbels eight or nine inches wide; the petioles had inflated sheaths."

At this point, Thoreau's Worcester friends, H. G. O. Blake and Theo Brown, "wet, ragged, and bloody with black flies," joined them. They had seen the smoke of Wentworth's fire.

"*July* 9. *Friday*. Walked to the Hermit Lake, some forty rods northeast." Here Thoreau recorded the plants he saw.

"I ascended the stream in the afternoon and got out of the ravine at its head, after dining on chiogenes [creeping snowberry, *Gaultheria hispidula*] tea, which plant I could gather without moving from my log seat. We liked it so well that Blake gathered a parcel to carry home."

A year earlier, the Indian guide Joe Polis had served the same drink to Hoar and Thoreau in Maine. Thoreau judged Tuckerman's Ravine to be between 1000' and 1500' deep, and he was right — the altitude gained between Hermit Lake and Tuckerman Junction is 1506'. This beautiful glacial cirque honors Edward Tuckerman, who, born with Thoreau in 1817, became professor of botany at Amherst College in 1858, the year of Thoreau's presence here, the year his name first appeared on maps, though it had been applied to the ravine for a decade. Thoreau used Tuckerman's books on lichens. This camp (3857–3877') was Thoreau's highest, just above the collier's shanty of 7 July, and Katahdin.

"Returning, I sprained my ankle in jumping down the brook, so that I could not sleep that night, nor walk the next day....

"The black flies, which pestered us till into evening, were of various sizes, the largest more than an eighth of an inch long. There were scarcely any mosquitoes here, it was so cool.

"A small owl came in the evening and sat within twelve feet of us, turning its head this way and that and peering at us inquisitively. It was apparently a screech owl." Unsure, he added: "Or *Acadica??* Saw-whet?" Acadica or Acadian, from the Latin species name of the saw-whet (*Aegolius acadicus*), referring to the place where it was first collected and named, Acadia or Nova Scotia, is now obsolete as a common name. "Saw-whet owl is the only species that would be probable," Mark Suomala of the New Hampshire Audubon Society told me, noting, "But nothing is impossible! Eastern screech owls are usually only found further south today."

July 10. "The only animals we saw about our camp were a few red squirrels.... The *Fringilla hyemalis* [slate-colored junco, *Junco hyemalis,* also snowbird to Thoreau] was most common in the upper part of the ravine, and I saw a large bird of prey, perhaps an eagle, sailing over the head of the ravine. The wood thrush and veery sang regularly, especially morning and evening.

But, above all, the peculiar and memorable songster was that Monadnock-like one, keeping up an exceedingly brisk and lively strain. It was remarkable for its incessant twittering flow.... It reminded me of a fine corkscrew stream issuing with incessant lisping tinkle from a cork, flowing rapidly, and I said that he had pulled out the spile and left it running."

This songster, unseen while singing, Thoreau could not identify. Francis H. Allen, an ornithologist and a Thoreau devotee, believed the voice was that of a winter wren. Mark Suomala agreed. Allen also said that Thoreau's wood thrush and veery were the olive-backed thrush (Swainson's) and Bicknell's thrush (gray-cheeked).

Sunday night's rain did not dampen Thoreau's party. On Monday, 12 July, Thoreau's forty-first birthday, they walked out "by a path about two and a half or three miles to carriage-road." This is the Raymond Path, which joins the auto road just below the 2-mile mark. On the same road they had ascended five days before, they went down to the Glen House. "Dined by Peabody River, three quarters of a mile south of Glen House....

"In the afternoon we rode along, three of us, northward and northwestward on our way round the mountains, going through Gorham. We camped about a mile and a half west of Gorham, by the roadside, on the bank of Moose River." Apparently, Wentworth had gone home. Did Theo Brown, too?

"*July* 13. *Tuesday.* This morning it rained, keeping us in camp till near noon, for we did not wish to lose the view of the mountains as we rode along.

"We dined at Wood's tavern in Randolph, just over Randolph Hill, and here had a pretty good view of Madison and Jefferson...but a cloud rested on the summits most of the time....

"It rained steadily and soakingly the rest of the afternoon....

"We put up at a store just opposite the town hall on Jefferson Hill. It here cleared up at sunset, after two days' rain, and we had a fine view of the mountains, repaying us for our journey and wetting, Mt. Washington being some thirteen miles distant southeasterly. Southwestward we looked down over a very extensive, uninterrupted, and level-looking forest.... By going still higher up the hill, in the wet grass north of the town house, we could see the whole White Mountain range from Madison to Lafayette....

"After the sun set to us, the bare summits were of a delicate rosaceous color, passing through violet into the deep dark-blue or purple of the night which already invested their lower parts, for this night-shadow was wonderfully blue, reminding me of the blue shadows on snow. There was an afterglow in which these tints and variations were repeated. It was the grandest mountain view I ever got."

Thoreau went from Gorham to Jefferson along what is now Route 2. Wood's Tavern, formerly on the north side of Route 2 east of Old Pinkham Notch Road, is today Broad Acres Farm. Jefferson Hill is now Jefferson. The town hall stood along Waumbek Brook, where the road to Mount Starr King Trail is now; thus Thoreau's view was from the south side of this mountain. His lodging, known in later years as Cherry Cottage, was removed in 1972. From the turnout east of Jefferson, the prospect is outstanding.

Lafayette

"*July* 14. *Wednesday*. This forenoon we rode on through Whitefield to Bethlehem.... Climbed the long hill from Franconia to the Notch, passed the Profile House, and camped half a mile up the side of Lafayette....

"*July* 15. *Thursday*. Continued the ascent of Lafayette [5260'] also called the Great Haystack. It is perhaps three and a half miles from the road to the top by path along winding ridge." The Old Bridle Path runs 2.9 miles from Franconia Notch Parkway to the AMC Greenleaf hut, where Greenleaf Trail continues 1.1 miles to the summit. It was adjacent to the hut where Thoreau inspected "a little pond," Eagle Lake (c. 4200'). "In the dwarf fir thickets above and below this pond, I saw the most beautiful linnaeas [twinflower, *Linnaea borealis*] that I ever saw. They grew quite densely, full of rose-purple flowers,— deeper reddish-purple than ours, which are pale,—perhaps nodding over the brink of a spring, altogether the fairest mountain flowers I saw....

"We had fine weather on this mountain, and from the summit a good view of Mt. Washington and the rest, though it was a little hazy in the horizon. It was a wild mountain and forest scene from south-southeast round easterwardly to north-northeast. On the northwest the country was half cleared, as from Monadnock,—the leopard-spotted land."

They returned to the "little pond" for dinner, and Thoreau busied himself counting fir and spruce rings.

"When half-way down the mountain, amid the spruce, we saw two pine grosbeaks, male and female, close by the path, and looked for a nest, but in vain. They were remarkably tame, and the male a brilliant red orange,— neck, head, breast beneath, and rump,—blackish wings and tail, with two white bars on wings. (Female, yellowish.) The male flew nearer inquisitively, uttering a low twitter, and perched fearlessly within four feet of us, eying us and pluming himself and plucking and eating the leaves of the *Amelanchier oligocarpa* [Bartram shadbush, *A. bartramiana*] on which he sat, for several minutes. The female, meanwhile, was a rod off. They were evidently breeding there. Yet neither Wilson nor Nuttall speak of their breeding in the United States." Thoreau used the ornithologies of the naturalists Alexander Wilson and Thomas Nuttall; Nuttall had taught at Harvard before Thoreau went there.

"Rode on and stopped at Morrison's (once Tilton's) Inn in West Thornton." Thoreau had thus retraced his first journey through Franconia Notch in the reverse direction and rested again where he and his brother had on 6 September 1839.

On 16 July, traveling south through Thornton and Campton, Thoreau mused, "Indeed the summit of a mountain, though it may appear thus regular at a distance, is not, after all, the easiest thing to find, even in clear weather. . . .

"Cannon Mountain [4100'] on the west side of the Franconia Notch (on whose side is the profile [the Old Man of the Mountain]) is the most singularly lumpish mass of any mountain I ever saw, especially so high. It looks like a behemoth or a load of hay, and suggests no such pyramid as I have described.... I would say that the eye needs only a hint of the general form and completes the outline from the slightest suggestion....

"Nooned on west bank of the Pemigewasset, half a mile above the New Hampton covered bridge....

"Lodged at tavern in Franklin, west side of river.

"*July* 17. *Saturday.* Passed by Webster's place, three miles this side of the village."

Daniel Webster once went to see Thoreau's aunt Louisa Dunbar, who lived with the Thoreaus. Webster's birthplace, a state historic site, is a few miles southwest of West Franklin via Route 127. There is a half-mile nature trail at the site.

"Spent the noon on the bank of the Contoocook in the northwest corner of Concord, there a stagnant river owing to dams....

"Reached Weare and put up at a quiet and agreeable house, without any sign or barroom....

"*July* 18. *Sunday.* Keep on through New Boston, the east side of Mount Vernon, Amherst to Hollis, and noon by a mill-pond in the woods, on Pennichook Brook [Pennichuck], in Hollis, or three miles north of village. At evening go on to Pepperell."

Entering Amherst on the Boston Post Road, Thoreau must have recalled his sleigh ride to Amherst from Nashua on the "bitter cold afternoon" of 18 December 1856, bundled in a buffalo robe. A splendid historical tour around this common took me six hours! I saw the Congregational church where Thoreau "lectured in basement (vestry) of the orthodox church, and I trust helped to undermine it." His hotel is gone. Five miles north of Amherst Common, on the Bedford line, Thoreau's friend Horace Greeley, editor of the *New York Tri-*

bune, was born. His farmhouse, a restored, augmented Cape, identified by stone marker on Horace Greeley Road, occupied a pastoral setting when I visited in 1985, though large new homes were encroaching.

Thoreau's last reflection of his Mount Washington trip spoke for the wayfaring public. "What barbarians we are! The convenience of the traveller is very little consulted. He merely has the privilege of crossing somebody's farm by a particular narrow and maybe unpleasant path. The individual retains all other rights,—as to trees and fruit, and wash of the road, etc. On the other hand, these should belong to mankind inalienably. The road should be of ample width and adorned with trees expressly for the use of the traveller. There should be broad recesses in it, especially at springs and watering-places, where he can turn out and rest, or camp if he will."

Arriving home the next noon, 19 July, he listed by zone the plants he had seen, the first such description for Mount Washington and a significant contribution to alpine ecology.

"I noticed, when I was at the White *Mts* last, several nuisances which render travelling there-abouts unpleasant," Thoreau wrote H. G. O. Blake on 4 November 1860. "The chief of these was the *mt* houses. I might have supposed that the main attraction of that region even to citizens, lay in its wildness and unlikeness to the city, & yet they make it as much like the city as they can afford to. I heard that the Crawford House was lighted with gas, & had a large saloon, with its band of music, for dancing. But give me a spruce house made in the rain" (*C,* 596–97).

TEXT: *The Journal of Henry D. Thoreau,* XI:3–62.

Sources and Further Reading

Text Sources of Thoreau's Writings

The Correspondence of Henry David Thoreau. Walter Harding and Carl Bode, eds. New York University Press, 1958; Greenwood Press, 1974.

The Journal of Henry D. Thoreau. Bradford Torrey and Francis H. Allen, eds. Houghton Mifflin, 1906; Dover Publications, 1962.

Journal, Volume 1: 1837–1844. Elizabeth Hall Witherell, William L. Howarth, Robert Sattelmeyer, and Thomas Blanding, eds. Princeton University Press, 1981.

Journal, Volume 2: 1842–1848. Robert Sattelmeyer, ed. Princeton University Press, 1984.

Journal, Volume 3: 1848–1851. Robert Sattelmeyer, Mark R. Patterson, and William Rossi, eds. Princeton University Press, 1990.

Journal, Volume 4: 1851–1852. Leonard N. Neufeldt and Nancy Craig Simmons, eds. Princeton University Press, 1992.

Journal, Volume 5: 1852–1853. Patrick F. O'Connell, ed. Princeton University Press, 1997.

The Maine Woods. Joseph J. Moldenhauer, ed. Princeton
    University Press, 1972.

Walden. J. Lyndon Shanley, ed. Princeton University Press,
    1971.

"A Walk to Wachusett" (*Boston Miscellany,* January 1843)
    and "Walking" (*Atlantic Monthly,* June 1862). In
    Henry D. Thoreau, *Excursions,* Ticknor and Fields,
    1863; and *Henry David Thoreau: The Natural His-*
    tory Essays, introduction and notes by Robert Sat-
    telmeyer, Peregrine Smith Books, 1980. Selections are
    taken from the latter source.

A Week on the Concord and Merrimack Rivers. Carl. F.
    Hovde, William L. Howarth, and Elizabeth Hall
    Witherell, eds. Princeton University Press, 1980.

A Yankee in Canada. Ticknor and Fields, 1866; Harvest
    House, 1961.

Short Titles Used in the Text

C   *The Correspondence of Henry David Thoreau.*
        Walter Harding and Carl Bode, eds. New York
        University Press, 1958; Greenwood Press, 1974.

JMN   *The Journals and Miscellaneous Notebooks*
        of Ralph Waldo Emerson. William H. Gilman
        et al., eds. 16 vols. Harvard University Press,
        1960–1982.

Letters *The Letters of Ralph Waldo Emerson.* Ralph L.
        Rusk, ed. 6 vols. Columbia University Press,
        1939. Continued under the editorship of Eleanor
        M. Tilton, 4 vols. (7 to 10). Columbia University
        Press, 1990–1995.

Log   Borst, Raymond R. *The Thoreau Log: A Docu-*

mentary Life of Henry David Thoreau,
1817–1862. G. K. Hall, 1992.

PN   *The Poetry Notebooks of Ralph Waldo Emerson.*
Ralph H. Orth, et al., eds. University of Missouri
Press, 1986.

Further Reading

Allison, Elliott. "Thoreau in Vermont," *Vermont Life,* Autumn 1954, 11–13. Draws from Cabot's *Annals of Brattleboro.* Thoreau was not a guest in the house (10 Chase Street) shown in the accompanying photograph. I am indebted to Harriet Ives for the correct place (Land Records Book 28:194, in town clerk's office).

Allison, Elliott. "Thoreau of Monadnock," *Thoreau Journal Quarterly,* October 1973, 15–21. Channing did not accompany Thoreau in 1852.

Allison, Elliott. "Alone on the Mountain," *Yankee,* June 1973, 158–165. William Ellery Channing climbed Monadnock on several occasions. His journal of 1870 (30 July–16 August) and 1871 (14–15 August) and his reflections of November 1875 are reprinted here. The original is in the Houghton Library, Harvard University.

Allison, Elliott. "Mark Twain in Dublin," *New Hampshire Profiles,* August 1967, 28–32. Twain stayed in the Henry Copley Greene House on Lone Tree Hill from May to November 1905, writing two short stories and *Eve's Diary.* He returned in May 1906 for four months, this time to Upton's Mountain View Farm, where he dictated his autobiography while his biographer, Albert Bigelow Paine, drove out mornings from a rented room in Dublin, often walking out in afternoons as well. See his *Mark Twain: A Biography* (1912; Chelsea House, 1980) for Dublin, 3:1237–1248, 1307–1314. I am indebted to Elliott Allison for showing me these houses, which his father had located for Twain.

Allison, Elliott. "A Thoreauvian on Red Hill," *Yankee,* June 1950, 36–38, 43, 75–76; "Thoreau's Disciple on Red Hill," *Appalachia,* June 1964, 116–121.

Angelo, Ray. *Botanical Index to the Journal of Henry David Thoreau.* Gibbs M. Smith, 1984. He is my authority on botanical names.

Bartram, William. *Travels Through North and South Carolina* (1791). In *Travels and Other Writings,* edited by Thomas P. Slaughter, Library of America, 1996. Thoreau read this book, wherein the Catskills made their literary debut, Alf Evers says. In September 1753, John Bartram brought his son William to North and South lakes to collect balsam fir seeds for British gardens.

Bode, Carl, ed. *Collected Poems of Henry Thoreau.* Johns Hopkins University Press, 1964. Thoreau included his poems "With Frontier Strength Ye Stand Your Ground" and "Not Unconcerned Wachusett Rears His Head" in his Wachusett essay.

Burns, Deborah E., and Lauren R. Stevens. *Most Excellent Majesty: A History of Mount Greylock.* Berkshire County Land Trust and Conservation Fund, 1988. Excerpts from Thoreau, 93–100. Lithograph of North Adams in 1841 shows extensive open land on mountain. Excellent list of sources.

The Writings of John Burroughs. 23 vols. Houghton Mifflin, 1904ff; Russell and Russell, 1968. The nineteenth-

century literary naturalist of the Catskills John Burroughs (1837–1921) was born near Roxbury, where
Burroughs Memorial Field and Woodchuck Lodge,
the naturalist's grave and his summer abode from
1910, respectively, sit adjacent to each other. He later
lived at Riverby (now a private home) in West Park,
New York, where he wrote in a cabin, Slabsides, a
mile west. John Muir visited Slabsides on 22 June
1896. Though Slabsides is open only two days a year,
you are welcome to walk the grounds year-round.
Burroughs wrote about Emerson, Whitman, and
Thoreau. For Thoreau see, especially, *Indoor Studies*
(1889), 3–47; *Literary Values* (1902), 217–223, and *The
Last Harvest* (1922), 103–171.

Burt, Allen F. *The Story of Mount Washington.* Hanover,
N.H.: Dartmouth Publications, 1960. This classic
tells of visits by Thoreau (46–47), Emerson (43),
Hawthorne (43) and Louisa May Alcott (79–80).

Cabot, Mary R. *Annals of Brattleboro, 1681–1895.* Brattle-
boro: E. L. Hildreth, 1921. 2 vols. Addison and Ann
Brown, 1:394–398; Charles C. Frost, 1:476–479;
Thoreau, 1:323; Wesselhelft Water-Cure, 2:563–586;
Brattleboro Retreat, 1:424–436.

Capper, Charles. *Margaret Fuller: An American Romantic
Life: The Private Years.* Oxford University Press. I am
indebted to Professor Capper for the Ben Lomond
story, in *Margaret Fuller: "These Sad but Glorious
Days": Dispatches from Europe, 1846–1850,* edited by
Larry J. Reynolds and Susan Belasco Smith, Yale
University Press, 1991, 69–77. Capper's second volume, of Fuller's life from 1840 to 1850, is eagerly
awaited.

Carr, Gerald L. *Frederic Edwin Church: Catalogue Rai-sonné of Works of Art at Olana State Historic Site.* 2 vols. Cambridge University Press, 1994. In summer 1852, Church made his first visit to Kineo by steamer on Moosehead Lake and to Katahdin. He returned to both in 1856, and to Katahdin in 1876, 1877, 1878, 1879, 1881. His last dated work was *Mount Katahdin from Millinocket Camp,* 1895. There is still no evidence that Church read Thoreau's "Ktaadn," which first appeared serially in *Union Magazine of Literature and Art* in 1848.

Chamberlain, Allen. *The Annals of the Grand Monadnock.* 1936; Society for the Protection of New Hampshire Forests, 1975, 70–80. Chamberlain believed Thoreau used the now nonexistent Eveleth's path to the ridge, 57. For Eveleth, 135. The original edition contains a helpful map that locates Thoreau's landmarks. Elliott Allison told me that Chamberlain spent his summers at the Ark (extant) in Jaffrey.

Champney, Benjamin. *Sixty Years' Memories of Art and Artists.* Woburn, Mass., 1900. " 'Beauty Caught and Kept': Benjamin Champney in the White Mountains," *Historical New Hampshire,* Fall–Winter 1996. Exhibition catalogue with essays by Charles O. Vogel and Donald D. Keyes.

Corrigan, Rupert E. *Jefferson, New Hampshire, Before 1996.* Town of Jefferson, 1995. Thoreau stayed in Cherry Cottage, 109–110.

Dwight, Timothy. *Travels in New England and New York.* Edited by Barbara Miller Solomon with Patricia M. King. 4 vols. Harvard University Press, 1969.

Emerson, Edward W. "The Grand Monadnock," *New Eng-*

land Magazine, September 1896, 33–51. Frontispiece is a painting of Monadnock by Edward Emerson, who, inspired by Thoreau, made his first ascent of Monadnock in July 1861. Edward and his sister Ellen regularly camped on the mountain. In June 1866, they were joined by their father, Ralph Waldo Emerson, and by William Ellery Channing and other friends. An unrelenting storm drove Waldo with four reluctant women to the Mountain House for the night, leaving space in the tent for the others. One of the party, Annie Keyes, married Edward on 19 September 1874. They summered at Monadnock with their family, building a home there in 1897 that still looks northward to the mountain from Route 124 west of the Old Toll Road. In 1918 Edward guided photographer Herbert Gleason to the site of Thoreau's 1858 camp.

Emerson, Ralph Waldo. "Monadnoc." In *Ralph Waldo Emerson: Collected Poems and Translations* (Library of America, 1994), 49–60. Gay Wilson Allen, in *Waldo Emerson* (1981; Penguin Books, 1982), 486–487, says Emerson wrote verses on Monadnock on 3 May 1845, none of which appeared in the final poem. However, Professor Albert J. von Frank informs me that on that date Emerson was on Mount Wachusett (*PN,* 795, 861–862).

Evers, Alf. *The Catskills: From Wilderness to Woodstock.* Garden City, N.Y.: Doubleday, 1972; New York: Overlook Press, 1984, 488–491. Evers also cites Thoreau's comparison of Walden with North and South lakes from "The First Version of Walden," in J. Lyndon Shanley, *The Making of Walden* (University of

Chicago Press, 1957; Midway reprint, 1973), 138–139.

Frizzell, Martha McDanolds. *A History of Walpole, New Hampshire.* 2 vols. Walpole Historical Society, 1963. Fall Mountain, 1:407.

Fuller, Richard. "Visit to Wachusett, July 1842," edited by Walter Harding, *Thoreau Society Bulletin,* Fall 1972, 1–4. This fragment from Fuller's notebook in the Boston Public Library gives little geography.

Graustein, Jeannette E. "Thoreau's Packer on Mt. Washington...," *Appalachia,* June 1957, 414–417. Evidence that Thoreau's packer was William H. H. Wentworth, not Lowell M. Wentworth, as Christopher McKee identifies (see p. 101).

Harding, Walter. "Thoreau in the White Mountains," *Coos Magazine* (Colebrook, N.H.), June 1993, 17. Originally appearing in the *New Hampshire Troubador,* July 1941, this was Harding's first published article on Thoreau.

Harding, Walter. *The Days of Henry Thoreau.* Knopf, 1965; Princeton University Press, 1992 (restores all footnotes and adds a new afterword).

Harding, Walter, ed. "Mount Washington in 1850: [Diary of] Franklin Benjamin Sanborn," *Appalachia,* June 1952, 17–20. His ascent, on 15 September, from Horace Fabyan's hotel to the summit, he figured at 12 miles (I'd estimate 9) in 5 hours, descending from 4 to 8 P.M. Undoubtedly, his route was the second Crawford Path, built in 1821 by Ethan Allen Crawford and Charles Jesse Stuart, which went from Crawford's to the base of Ammonoosuc Ravine and from there up its north shoulder, which Fabyan in the 1840s made into a bridle path. Sanborn went on to Thomas Crawford's (the Notch House), expecting to stay the night. Thomas Crawford, a son of Abel Crawford (not a nephew, as Sanborn states), was closing for the season, however, so Sanborn accompanied him to Littleton. Sanborn lived in Concord, Massachusetts, from 1855 until his death in 1917, the last of the transcendentalists to die, Harding says. He wrote *Henry D. Thoreau* (Houghton Mifflin, 1882).

Harding, Walter, ed. *The Collected Poems of William Ellery Channing the Younger, 1817–1901.* Scholars' Facsimiles and Reprints, 1967. "Wachusett," 240–260. For criticism of this poem, see Lawrence Buell, *Literary Transcendentalism: Style and Vision in the American Renaissance* (Cornell University Press, 1973), 239–262.

Hawthorne, Nathaniel. *The American Notebooks.* Edited by Claude M. Simpson. Ohio State University Press, 1972, 79–149. Hawthorne's residency in North Adams, Massachusetts (26 July–11 September 1838), provided communion with Greylock, even a Williams College commencement (15 August), and walks anticipating Thoreau's along the Bellows Pipe Trail (22 August and 9 September). Of his stay here only one letter survives. *Nathaniel Hawthorne: The Letters, 1813–1843.* Edited by Thomas Woodson, L. Neal Smith, and Norman Holmes Pearson. Ohio State University Press, 1984, 274–275.

Hayes, Lyman Simpson. *History of the Town of Rockingham, Vermont...1753–1907.* Town of Bellows Falls, Vermont, 1907. Thoreau's view is conveyed in an 1855 lithograph of Bellows Falls from Table Rock.

Hedrick, Joan D. *Harriet Beecher Stowe: A Life.* Oxford

University Press, 1994. Chapter on the Water Cure.

Higginson, Thomas Wentworth. "Going to Mount Katahdin," *Putnam's Monthly,* September 1856, 242–255. Reprinted in *Appalachia,* June 1925, 101–129. See also his *Letters and Journals,* edited by Mary Thacher (Houghton Mifflin, 1921), 117–121. Thoreau wrote Higginson about itineraries and provisions, 28 January 1858 (*C,* 506–508).

Higginson, Thomas Wentworth, and Mary Thacher Higginson. *Such as They Are.* Roberts Brothers, 1893. Elliott Allison showed me this book, which Thomas Wentworth Higginson had autographed and given to his father. One of Mary's poems herein, "Glimpsewood," is of their home on the south shore of Dublin Lake, west of the Pumpelly trailhead. They bought an acre and a half in 1890 and built Glimpsewood the following year; it still stands. Thomas Wentworth Higginson (1823–1911), a protean soul, was Emily Dickinson's literary agent. After his wife, Mary Channing, died, he married Mary Thacher, who wrote a biography of him (1914). The Allisons often met Thomas Wentworth Higginson walking around Dublin Lake; Elliott was proud to have shaken hands with a man who had shaken Thoreau's hand.

Hoag, Ronald Wesley. "The Mark on the Wilderness: Thoreau's Contact with Ktaadn," *Texas Studies in Literature and Language,* Spring 1982, 23–46. Hoag argues that Thoreau experienced the sublime on Katahdin.

Howarth, William. *Thoreau in the Mountains: Writings by Henry David Thoreau.* Farrar, Straus and Giroux, 1982.

Huber, J. Parker. *The Wildest Country: A Guide to Thoreau's Maine.* Appalachian Mountain Club, 1981. See Further Reading sections for Katahdin and Kineo.

Jackson, Charles T. *Final Report on the Geology and Mineralogy of the State of New Hampshire.* Concord, N.H.: Carroll & Baker, 1844. From Red Hill (71–72), Jackson found "the most beautiful mountain view which this country affords."

Kaaterskill: From the Catskill Mountain House to the Hudson River School. 1993; Mountain Top Historical Society, Black Dome Press, 1996. A map from Van Loan's *Catskill Mountain Guide* (New York, 1897) locates Glen Mary Cottage and Scribner's mill, 50–51. Justine L. Hommel, "Kaaterskill Lodgings," speaks of Glen Mary Cottage, 88–89. I am indebted to historian Raymond Beecher for obituaries of Mary and Ira Scribner from the *Catskill Examiner,* 2 March 1889 and 20 September 1890, respectively.

Kelly, Franklin, with Stephen Jay Gould, James Anthony Ryan, and Debora Rindge. *Frederic Edwin Church.* Washington, D.C.: National Gallery of Art, 1989. Exhibition catalogue. Ryan's "Frederic Church's Olana: Architecture and Landscape as Art," 126–156.

Kipling, Rudyard. "In Sight of Monadnock," *Letters of Travel.* New York: Doubleday, Page, 1920, 3–16. I am indebted to the late Mrs. F. Cabot Holbrook of Brattleboro, owner of Naulakha before the Landmark Trust, for her tour.

Laski, J. K. "Dr. Young's Botanical Expedition to Mount Katahdin," *Maine Naturalist,* June 1927, 38–62. Reprinted from *Bangor Courier,* September 1847, which, it appears, Sophia sent her brother from Bangor (*C,* 187).

Ludington, Townsend. *Marsden Hartley: The Biography of an American Artist.* Little, Brown, 1992. In October 1939, Hartley (1877–1943) spent eight days at Katahdin Lake depicting, not climbing, the mountain. Back in Bangor, he read Thoreau's *Journals,* most likely in the Bangor Public Library, which has the 1906 edition, and produced some twenty mountain paintings in three years.

McGill, Frederick T., Jr. *Channing of Concord: A Life of William Ellery Channing II.* Rutgers University Press, 1967. Channing made several trips to the White Mountains. On his first, summer 1836, he walked through Crawford Notch, meeting the Crawfords. According to F. B. Sanborn's notes of 15 September 1901 in the Concord Free Public Library, Channing subsequently stayed at Tom Crawford's, climbing "the mountain just above," which is Willard (2850'), not Washington.

McKee, Christopher. "Thoreau: A Week on Mt. Washington and in Tuckerman Ravine," *Appalachia,* December 1954, 169–183. Distinguishes Thoreau's *Journal* account from that of his early biographers, Frank B. Sanborn and William Ellery Channing.

McKee, Christopher. "Thoreau's First Visit to the White Mountains," *Appalachia,* December 1956, 199–209.

McKee, Christopher. "Thoreau's Sister in the White Mountains," *Appalachia,* December 1957, 551–556. Sophia ascended Mount Washington on the cog railway on 9 August 1870, spent a sleepless night in Tip Top House, and went down the next day by the carriage road to the Glen House, where five hundred guests danced to a Boston band.

Moldenhauer, Joseph J. *"The Maine Woods,"* in *The Cam-*

bridge Companion to Henry David Thoreau, edited by Joel Myerson. Cambridge University Press, 1995, 124–141.

Murray, Stuart. *Rudyard Kipling in Vermont: Birthplace of the Jungle Books.* Bennington, Vt.: Images from the Past, 1997. Arriving in Brattleboro by train on 17 February 1892, Rudyard and Caroline Kipling sleighed over moonlit snow to her brother's farmhouse, from where, the next day, inspired by Emerson's poem "Monadnoc," the Kiplings climbed the hillside to see the mountain. They decided to build Naulakha there to behold Monadnock.

Myers, Kenneth. *The Catskills: Painters, Writers and Tourists in the Mountains, 1820–1895.* University Press of New England, 1988. This catalogue features paintings by Cole, Church (*Twilight Among the Mountains,* 1844), and Sanford Robinson Gifford. Gifford (1823–1880) was living in Hudson, New York, when Thoreau passed through (133); he stayed with Ira and Mary Scribner in late September 1869 (136).

Myerson, Joel, and Daniel Shealy. "Louisa May Alcott on Vacation: Four Uncollected Letters," *Resources for American Literary Study,* Spring–Autumn 1984, 113–141. This article reprints Alcott's letters to the *Boston Commonwealth* in 1863 of her visit to the White Mountains in July 1861. While staying for about a month at the Alpine House, opposite the depot in Gorham, she ascended Washington by carriage.

Myerson, Joel, Daniel Shealy, and Madeleine B. Stern, eds. *The Selected Letters of Louisa May Alcott.* Little, Brown, 1987; *The Journals of Louisa May Alcott.* Little, Brown, 1989, which supplant Cheney's edition.

Noble, Louis LeGrand. *The Life and Works of Thomas Cole.* Edited by Elliot S. Vesell. 1853; Harvard University Press, 1964. Noble was a close friend of Cole and Church. As a minister, he married Church and Isabel Carnes on 14 June 1860 in Dayton, Ohio, and told of his travels with Church in *After Icebergs with a Painter: A Summer Voyage to Labrador and around Newfoundland* (New York: D. Appleton, 1861).

Parker, Hershel. *Herman Melville: A Biography.* Volume 1: 1819–1851. Johns Hopkins University Press, 1996. This begins Melville's life in the Berkshires, which continues in volume 2 (forthcoming), wherein Parker discusses October Mountain and the claim that Melville wrote a story with that title.

Richardson, Robert D., Jr. *Henry Thoreau: A Life of the Mind.* University of California Press, 1986.

Rudolph, Frederick. *Mark Hopkins and the Log: Williams College, 1836–1872.* Yale University Press, 1956. Puritan Williams College upbraided transcendentalists— as does Rudolph, who says Thoreau "uttered grim pronouncements on the age and crawled into the seclusion of a cabin on Walden Pond" (36), and misdates his Berkshire visit (135).

Sattelmeyer, Robert. "A Walk to More than Wachusett," *Thoreau Society Bulletin,* Winter 1992, 1–4. Sesquicentennial talk in Princeton, Massachusetts, 8 July 1992, stresses the significance of "his first travel essay."

Sattelmeyer, Robert. *Thoreau's Reading: A Study in Intellectual History with Bibliographical Catalogue.* Princeton University Press, 1988.

Sinclair, Warren M. *Wachusett: Wajuset Gatherings from Then and When.* Salem, Mass.: Higginson Book Company, 1996. Contains Thoreau excerpts.

Smith, J. E. A. *History of Pittsfield (Berkshire County), Massachusetts.* Boston: Lee and Shepard, 1869–1876. 2 vols. The depot is shown on a map of Pittsfield. Berkshire Jubilee, 2:573–596.

Stebbins, Theodore E., Jr. "Thomas Cole at Crawford Notch," *National Gallery of Art: Report and Studies in the History of Art,* 1968, 133–145.

Swift, Esther Munroe, and Mona Beach. *Brattleboro Retreat, 1834–1984: 150 Years of Caring.* Brattleboro Retreat, 1984. Thoreau is included among the famous who came to Brattleboro for the Water Cure (24), which is not correct. The first mountain land purchase is 1852 (138), while *Annals of Brattleboro* gives 1849 (1:433).

Torrey, Bradford. *Footing it in Franconia.* Houghton Mifflin, 1901. An editor of the 1906 Houghton Mifflin edition of Thoreau's *Journal,* Torrey loved searching for birds in Franconia Notch.

Waterman, Laura and Guy. *Forest and Crag: A History of Hiking, Trail Blazing, and Adventure in the Northeast Mountains.* Appalachian Mountain Club, 1989.

Welch, Sarah N. *A History of Franconia, New Hampshire.* Town of Franconia, 1972. Walking to Franconia from Woodstock on 26 September 1984, I searched for possible places for the Thoreaus. A mile south of the hamlet on Route 116, then .4 mile west, brought me to the house (c. 1859) on Ridge Road, with a glorious view of the Franconia Range, where Robert Frost lived from 1915 to 1920 (and most summers thereafter until 1938), now open to the public from Memorial Day to Columbus Day. Returning to 116 and taking that south for 1.3 miles, I came to the Franconia Inn (built by Zebedee Applebee in 1791; burned 1934 and rebuilt), opposite the airport. At dark, .4 mile farther south, I found the Cannon Mountain House (c. 1850s), owned by Jerry Kosch, who had been there fifty years. The next day I began my walk to Crawford Notch, going down lovely Wells Road to the Horse and Hound Inn (c. 1830). At Route 18 I turned north and stopped at Lovett's Inn, just north of Route 141. The owner, Charles Lovett, Jr., told me that his father opened this farmhouse (c. 1784–1799) for guests in 1928. I then spoke with native Sarah Welch, whose great-grandfather, Luke Brooks, came here in 1795 and discovered the Old Man of the Mountain in 1805. Since Welch did not know of any inns in Franconia in 1839, I concluded that the Thoreaus found lodging in a residence.

Woodbury, Charles J. *Talks with Ralph Waldo Emerson*. New York: Baker & Taylor, 1890. Refers briefly to Emerson and Greylock (15, 44). Woodbury published some of this in *Century*, February 1890, 621–627. John McAleer discusses Woodbury in *Ralph Waldo Emerson: Days of Encounter* (Little, Brown, 1984), 576–581.

Woodson, Thomas. "Thoreau's Excursion to the Berkshires and Catskills," *ESQ*, 2nd quarter 1975, 82–92. Still the best account of this excursion.

Acknowledgments

I am grateful for what I am & have.
My thanksgiving is perpetual.
Thoreau to H. G. O. Blake, 6 December 1856

A wealth of literature on Thoreau made this book possible. As a result of this project, my respect and love for these scholars have deepened. We are all indebted to them for what they have given us. I also wish to thank the following: Scott Alessio, archivist, Berkshire County Historical Society; Ray Angelo, botanist, Cambridge, Massachusetts; Ida Bagley, Gorham Public Library, New Hampshire; Peter Baldwin, Pittsfield, New Hampshire; Isabel Beal, Groton Historical Society, Massachusetts; Raymond Beecher, librarian and historian, Greene County Historical Society, Coxsackie, New York; Charles E. Beveridge, series editor, Frederick Law Olmsted Papers, Washington, D.C.; Sylvia Kennick Brown, archivist, Williams College; Christine M. Burchstead, Rockingham Public Library, Bellows Falls, Vermont; Charles Capper, professor of history, University of North Carolina at Chapel Hill; Jerry Carbone, director, Richard Shuldiner, reference librarian, and the staff of Brooks Memorial Library, Brattleboro, Vermont; Sherman Clebnik, professor of geology, Eastern Connecticut State University; William Copeley, librarian, New Hampshire Historical Society; Rupert E. Corrigan, Jefferson,

New Hampshire; Curt Crow, National Geodetic Survey, Massachusetts and New Hampshire; Ellen S. Derby, Peterborough Historical Society, New Hampshire; Elaine Dixon, Putney, Vermont; Ann Young Doak, Williamsburg, Virginia; Albert J. von Frank, professor of English and American studies, Washington State University; Julie Friedman, Franconia, New Hampshire; Justine L. Hommel, historian, Town of Hunter, New York; Harriet Ives, Brattleboro Historical Society, Vermont; Gail LaGess, librarian, Florida Free Library, Massachusetts; John Lancaster, curator of special collections, and the staff of Amherst College Library; Robert Popp, Nongame and Natural Heritage Program, Vermont Department of Fish and Wildlife; Louise Pryor, Hampstead Public Library, New Hampshire; Warren Quinn, Temple, New Hampshire; Mark Suomala, New Hampshire Audubon Society; Jennie Swartz, USGS; Charles O. Vogel, Townsend, Massachusetts; Tom Wessels, Antioch New England Graduate School, Keene, New Hampshire; Cynthia S. Williams, curatorial assistant, Olana State Historic Site, Hudson, New York; Melanie Wisner, Houghton Library, Harvard University; Joyce T. Woodman, Concord Free Public Library, Massachusetts.

A community of nature writers also invigorated my efforts, one of whom suggested that this book be as "the pure peace of giving/one's gold away" (Mary Oliver, "Goldenrod"). Another, Edward Lueders in *The Clam Lake Papers*, reminded me to honor "co-authorship with the gods." I've gone to all these mountains with Thoreau, whose spirit graces my life. My prayer is that you will discover, as Thoreau did, that there is an abundance of light in the world.

The Spirit of Thoreau

"How many a man has dated a new era in his life from the reading of a book," wrote Henry David Thoreau in *Walden*. Today that book, perhaps more than any other American work, continues to provoke, inspire, and change lives all over the world, and each rereading is fresh and challenging. Yet as Thoreau's countless admirers know, there is more to the man than *Walden*. An engineer, poet, teacher, naturalist, lecturer, and political activist, he truly had several lives to lead, and each one speaks forcefully to us today.

The Spirit of Thoreau introduces the thoughts of a great writer on a variety of important topics, some that we readily associate him with, some that may be surprising. Each book includes selections from his familiar published works as well as from less well known and even previously unpublished lectures, letters, and journal entries. Thoreau claimed that "to read well, that is, to read true books in a true spirit, is a noble exercise, and one that will task the reader more than any exercise which the customs of the day esteem." The volume editors and the Thoreau Society believe that you will find these new aspects of Thoreau an exciting "exercise" indeed.

This Thoreau Society series reunites Henry Thoreau with

his historic publisher. For more than a hundred years, the venerable publishing firm of Houghton Mifflin has been associated with standard editions of the works of Emerson and Thoreau and with important bibliographical and interpretive studies of the New England transcendentalists. Until Princeton University Press began issuing new critical texts in *The Writings of Henry D. Thoreau,* beginning with *Walden* in 1971, Thoreauvians were well served by Houghton Mifflin's twenty-volume *Walden* or manuscript edition of *The Writings of Henry David Thoreau* (1906). Having also published Walter Harding's annotated edition of *Walden* (1995), Houghton Mifflin is again in the forefront of Thoreau studies.

You are invited to continue exploring Thoreau by joining our society. For well over fifty years we have presented publications, annual gatherings, and other programs to further the appreciation of Thoreau's thought and writings. And now we have embarked on a bold new venture. In partnership with the Walden Woods Project, the Thoreau Society has formed the Thoreau Institute, a research and educational center housing the world's greatest collection of materials by and about Thoreau. In ways that the author of *Walden* could not have imagined, his message is still changing lives in a brand-new era.

For membership information, write to the Thoreau Society, 44 Baker Farm, Lincoln, MA 01773-3004, or call 781-259-4750. To learn more about the Thoreau Institute, write to the same address; call 781-259-4700; or visit the Web site:

www.walden.org.

WESLEY T. MOTT
Series Editor
The Thoreau Society